NIST Technical Note 1603

Experimental Study of the Effects of Fuel Type, Fuel Distribution, and Vent Size on Full-Scale Underventilated Compartment Fires in an ISO 9705 Room

Andrew Lock
Matthew Bundy
Erik L. Johnsson
Anthony Hamins
Gwon Hyun Ko
Cheolhong Hwang
Paul Fuss
Richard Harris

U.S. Department of Commerce
Building and Fire Research Laboratory
National Institute of Standards and Technology
100 Bureau Drive
Gaithersburg, MD 20899

October 2008

U.S. Department of Commerce
Carlos M. Gutierrez, Secretary
National Institute of Standards and Technology
James Turner, Acting Director

DISCLAIMER

Certain companies and commercial properties are identified in this paper in order to specify adequately the source of information or of equipment used. Such identification does not imply endorsement or recommendation by the National Institute of Standards and Technology, nor does it imply that this source or equipment is the best available for the purpose.

TABLE OF CONTENTS

1 Introduction 1
   1.1 Motivation and Objective 1
   1.2 Previous Work 2
   1.3 Experimental Scope 4

2 Experimental Design 6
   2.1 Design of the room 6
      2.1.1 Dimensions 6
      2.1.2 Materials 6
      2.1.3 Doorway Dimensions 9
      2.1.4 The Burners 9
   2.2 Overview of equipment 11
      2.2.1 Heat Release Rate Measurement 11
      2.2.2 Gas analyzers 13
      2.2.3 Gas Chromatography 16
         2.2.3.1 Gas Sample Storage System 19
      2.2.4 Soot Samples 21
         2.2.4.1 Gravimetric 21
         2.2.4.2 Real time extractive 23
      2.2.5 Thermocouples 25
         2.2.5.1 Aspirated Thermocouples 25
         2.2.5.2 Radiation Effects on Bare Beads 28
      2.2.6 Heat flux gauges 29
   2.3 Sampling locations 31
   2.4 Data acquisition 33
   2.5 Data post-processing 33
   2.6 Uncertainty 35

3 Results 37
   3.1 List of Test Conditions 37
   3.2 Heat Release Rate 40
   3.3 Temperatures 50
   3.4 Heat Flux 60

    3.5   Interior Compartment Gas Species 65

    3.6   Soot 73

        3.6.1   Gravimetric 73

        3.6.2   Real time extractive 75

4   Compartment chemistry analysis 78

    4.1   Mixture Fraction Analysis 78

        4.1.1   Definition of Mixture Fraction 79

        4.1.2   Mixture Fraction Uncertainty 82

        4.1.3   Species Composition Results in terms of Mixture Fraction 84

        4.1.4   Condensed-Phase Hydrocarbon Fuels 87

    4.2   Post-Compartment Product Yields 95

    4.3   Carbon Balance 97

    4.4   Combustion Efficiency 104

5   Scaling from RSE to FSE 110

6   Discussion 114

    6.1   Behavior of different fuels 114

        6.1.1   Liquid Fuels 114

        6.1.2   Solid Fuels 120

        6.1.3   Comparison 125

    6.2   Effect of fuel distributions 125

    6.3   Ventilation effects 130

    6.4   Heat release rate ramp 135

7   Summary 137

8   Future Work 139

9   References 140

10   Acknowledgements 143

A.   Channel Lists 145

B.   Equipment List 147

C.   Micro-GC Method Report 148

LIST OF FIGURES

Figure 2.1 Internal dimensions of ISO 9705 enclosure used in these experiments including multiple door widths and gas sample and temperature probe locations. All dimensions have an uncertainty of ± 2 cm. 7

Figure 2.2 Photograph of the actual ISO 9705 room used for experiments. The structural construction of sheet steal on steel studs can be seen along with the internal surface covering of ceramic fiber blanket. 8

Figure 2.3 Ceramic insulation retainers used to secure the ceramic fiber blanket to the sheet steel walls. The actual retainer is shown (left) as well as its installed configuration (right). 8

Figure 2.4: Free-burn and spray burner pan construction and dimensions. The dimensions of 50 cm, 70.7 cm, and 100 cm are all internal burner dimensions. All burners had a lip height of 10 cm. 10

Figure 2.5: Positioning of free-burn pan burners. The burners of different size were placed at the geometric center of the floor and/or along the centerline against the back wall. Pans in either position were mounted on load cells to measure the mass loss (or gain in the case of the spray burner) to determine fuel loss rate. For the spray burner cases only a single 70.7 cm x 70.7 cm pan was used in the center of the floor to catch the fuel spray. Both the pump fed pool burner and the gravel filled natural gas burner were located at position 1 inside the room. 10

Figure 2.6: Schematic drawing of 6 m square hood and exhaust stack instrumented for calorimetry measurements. Taken from Ref. [29] 12

Figure 2.7: Exhaust gas sampling system used for heat release rate measurement. 13

Figure 2.8: Schematic drawing of gas sampling system. 15

Figure 2.9: Schematic diagram of gas sample storage system in position B-A. 20

Figure 2.10: Positions of the control valves for the gas sample storage system. 20

Figure 2.11: Schematic drawing of gravimetric soot sampling system. 22

Figure 2.12: Schematic of real time extractive soot measurement probe. 25

Figure 2.13: Detailed drawing of aspirated thermocouple using NACA design [34]. 27

Figure 2.14: Schematic drawing of aspirated thermocouple measurement hardware. 28

Figure 3.1: Heat release rate for test ISONG3 comparing the ideal heat release rate, as imposed by gas flow rate, by the red dashed line and the measured, by oxygen loss calorimetry, solid blue line. 39

Figure 3.2: Heat release rate for test ISOHept5 comparing the ideal heat release rate, as imposed by pump flow rate, by the red dashed line and the measured, by oxygen loss calorimetry, solid blue line. 39

Figure 3.3: Heat release rate for test ISOHept9 (Heptane) comparing the ideal heat release rate, as measured by the burner mass loss rate, by the red dashed line and the measured, by oxygen loss calorimetry, solid blue line. 43

Figure 3.4: Heat release rate for test ISOProp15 (Iso-Propanol) comparing the ideal heat release rate, as measured by the burner mass loss rate, by the red dashed line and the measured, by oxygen loss calorimetry, solid blue line. 43

Figure 3.5: Heat release rate for test ISOHeptD12 (Heptane) comparing the ideal heat release rate, as measured by the burner mass loss rate, by the red dashed line and the measured, by oxygen loss calorimetry, solid blue line. 44

Figure 3.6: Heat release rate for test ISOHeptD13 (Heptane) comparing the ideal heat release rate, as measured by the burner mass loss rate, by the red dashed line and the measured, by oxygen loss calorimetry, solid blue line. 44

Figure 3.7: Fire leaving the door of the ISO 9705 room during test ISOHeptD12. It was not possible to view the inside of the room during this test. 45

Figure 3.8: Heat release rate for test ISOStyrene17 (Polystyrene) comparing the ideal heat release rate, as measured by the burner mass loss rate, by the red dashed line and the measured, by oxygen loss calorimetry, solid blue line. The mass loss reading was lost during the experiment due to warping of the burner pan. The ideal HRR values were smoothed because of excessive signal noise. 45

Figure 3.9: Images of the burner warping and moving during test ISOStyrene17. The burner was observed to be as much as 20 cm off of the floor in one corner. The burner warped because of its large size, $1m^2$, and the excessive heat transfer to the burner in the room. This caused a loss of the mass loss measurement in this experiment. 46

Figure 3.10: Heat release rate for test ISOPPD18 (Polypropylene) comparing the ideal heat release rate, as measured by the burner mass loss rate, by the red dashed line and the measured, by oxygen loss calorimetry, solid blue line. The mass loss reading was lost during the experiment due to warping of the burner pan. The ideal HRR values were smoothed because of excessive signal noise. 46

Figure 3.11: Heat release rate for test ISOHept22 (heptane) comparing the ideal heat release rate, as measured by the spray burner pump flow rate, by the red dashed line and the measured, by oxygen loss calorimetry, solid blue line. 47

Figure 3.12: Heat release rate for test ISOHept27 (heptane) comparing the ideal heat release rate, as measured by the spray burner pump flow rate, by the red dashed line and the measured, by oxygen loss calorimetry, solid blue line. This test featured a linear increase in the fuel delivery rate to observe the effects of a HRR ramp. 47

Figure 3.13: steady state heat release results. The dashed line indicates ideal or complete burning. 48

Figure 3.14: Comparison between temperatures measured from bare bead and aspirated thermocouples at front sampling location as a function of time for test ISONG3. 52

Figure 3.15: Comparison of averaged temperatures measured from bare bead and aspirated thermocouples at front sample location for test ISONG3. 53

Figure 3.16: Comparisons of averaged temperature measured at front and rear thermocouple trees for test ISOHeptD12 and ISOHeptD13. 53

Figure 3.17: Histories of temperature at front and rear sampling locations for test ISOHept22 (heat release rate measured from calorimeter was included to show fire condition). 55

Figure 3.18: Histories of temperature at front thermocouple trees for test ISOHept9. 55

Figure 3.19: Histories of temperature at rear thermocouple trees for test ISOHept9. 56

Figure 3.20: Averaged temperatures as a function of heat release rate at front sample location for all fuels tested. 57

Figure 3.21: Averaged temperatures as a function of heat release rate at rear sample location for all fuels tested. 57

Figure 3.22: Comparison of heat flux measurements made in the ceiling for 1/4 width (20 cm) doorway heptanes fuel test ISOHept9. 62

Figure 3.23: Comparison of the heat flux gauges positioned in the floor for 1/4 doorway (20 cm) heptanes fuel case ISOHept9. 62

Figure 3.24: Photograph of the center (front) and rear burners immediately after fire test ISOPropD14. 63

Figure 3.25: Comparison of heat flux gauge measurements for heat flux gauges positioned in the floor for 1/4 doorway (20 cm) distributed fuel isopropanol case ISOProp14. 63

Figure 3.26: Comparison of heat flux gauge measurements for heat flux gauges positioned in the floor for 1/4 doorway (20 cm) distributed fuel isopropanol case ISOProp14. 64

Figure 3.27: Transient gas volume fraction s and soot mass fraction of test ISOHept9 (Heptane). 67

Figure 3.28: Transient gas volume fraction s and soot mass fraction of test ISOPP18 (Polypropylene). 67

Figure 3.29: Transient gas volume fractions and heat release rate of test ISOPP18 (Polypropylene). 68

Figure 3.30: Gas species volume fractions from GC analysis of front sample location in ISOHept27 at t=1375 s. 68

Figure 3.31: Steady state average oxygen volume fraction measurements at front sample probe location. 69

Figure 3.32: Steady state average oxygen volume fraction measurements at rear sample probe location. 69

Figure 3.33: Steady state average CO_2 volume fraction measurements at front sample probe location. 70

Figure 3.34: Steady state average CO_2 volume fraction measurements at rear sample probe location. 70

Figure 3.35: Steady state average CO volume fraction measurements at front sample probe location. 71

Figure 3.36: Steady state average CO volume fraction measurements at rear sample probe location. 71

Figure 3.37: Steady state average THC volume fraction measurements at front sample probe location. 72

Figure 3.38: Steady state average THC volume fraction measurements at rear sample probe location. 72

Figure 3.39: Steady state gravimetric soot mass fraction measurements at front sample probe location. 74

Figure 3.40: Steady state gravimetric soot mass fraction measurements at rear sample probe location. 74

Figure 3.41: Soot mass concentration and heat release rate during polystyrene fire test 21. 75

Figure 3.42: Comparison of the optical and gravimetric soot mass fraction at the rear of the compartment in test #21. 76

Figure 3.43: Soot mass concentration and heat release rate during heptane fire test 26. 77

Figure 3.44: Soot mass concentration and heat release rate during heptane fire test 28. 77

Figure 3.45: Comparison of the optical and gravimetric soot mass fraction at the rear of the compartment in test #28. 78

Figure 4.1: The equivalence ratio as a function of mixture fraction for nonpremixed flames burning methane and n-heptane. 83

Figure 4.2: The mass fraction vs. the mixture fraction calculated by the single-parameter mixture fraction model. 83

Figure 4.3: Mass fractions of front and rear compartment gas species for the natural gas fire tests #1-#3, and #32: (a) transient measurements, (b) time-averaged measurements as a function of mixture fraction without soot, and (c) time-averaged measurements as a function of mixture fraction including soot. 86

Figure 4.4. Mass fractions of front and rear compartment gas species for the heptane fire tests #5, #9, #12, #13, #19, #22-#26 and #28: (a) transient measurements, (b) time-averaged measurements as a function of mixture fraction without soot, and (c) time-averaged measurements as a function of mixture fraction including soot. 89

Figure 4.5: Mass fractions of front and rear compartment gas species for the toluene fire tests #20 and #29: (a) transient measurements, (b) time-averaged measurements as a function of mixture fraction without soot, and (c) time-averaged measurements as a function of mixture fraction including soot. 90

Figure 4.6: Mass fractions of front and rear compartment gas species for the polypropylene fire tests #11 and #18: (a) transient measurements, (b) time-averaged measurements as a function of mixture fraction without soot, and (c) time-averaged measurements as a function of mixture fraction including soot. 91

Figure 4.7: Mass fractions of front and rear compartment gas species for the polystyrene fire tests #16, #17 and #21: (a) transient measurements, (b) time-averaged measurements as a function of mixture fraction without soot, and (c) time-averaged measurements as a function of mixture fraction including soot. 92

Figure 4.8: Comparison of mixture fraction calculated with and without soot using the time-averaged species measurements when the HRR was quasi-steady................. 93

Figure 4.9: Mass fractions of front and rear compartment gas species for the iso-propanol fire tests #14, #15 and #30: (a) transient measurements, (b) time-averaged measurements as a function of mixture fraction without soot, and (c) time-averaged measurements as a function of mixture fraction including soot. 94

Figure 4.10: The CO_2 volume fraction, X_{CO_2}, in the exhaust stack as a function of the fire heat release rate during the periods when the HRR was quasi-steady for each of the fuels tested (DF indicates the doorway fraction of 80 cm).................... 96

Figure 4.11: The CO volume fraction, X_{CO}*, in the exhaust stack as a function of the fire heat release rate during the periods when the HRR was quasi-steady for each of the fuels tested. 96

Figure 4.12: The values of F_{CO} and F_{soot} as a function of the local equivalence ratio for the time averaged measurements during the period when the HRR was quasi-steady.. 100

Figure 4.13: The CO and soot yields as a function of the local equivalence ratio for the time averaged measurements during the period when the HRR was quasi-steady.. 102

Figure 4.14: The ratio of the CO to soot yield as a function of the local equivalence ratio during the period when the HRR was quasi-steady. 103

Figure 4.15: The CO yield as a function of the soot yield during the period when the HRR was quasi-steady. Also shown is a line representing the results of Koylu [2]. 103

Figure 4.16: The combustion efficiency in the exhaust stack as a function of the ideal heat release rate for natural gas and heptane fuels under the condition of full doorway size (DF=1.0). The indicated uncertainty includes the 14% uncertainty of the calorimetry used to make the measurements. 106

Figure 4.17: The combustion efficiency in the exhaust stack as a function of the ideal heat release rate for various fuels under the condition of DF=0.25. The indicated

Figure 4.18: The combustion efficiency in the exhaust stack as a function of the ideal heat release rate for various doorway sizes in heptane fires. The curve fit lines are for illustrative purposes only. 107

Figure 4.19: The local combustion efficiency at the front and rear sample locations as a function of the ideal heat release rate for various doorway sizes in heptane fires. The curve fit lines are for illustrative purposes only 107

Figure 4.20: The burning fraction inside compartment as a function of ideal heat release rate under the condition of DF=0.125 in heptane fires. The curve fit line is for illustrative purposes only. 108

Figure 5.1: Example normalized scaling quantities with $Q_d^* = 0.17$, $\phi = 0.1$, and $H/D^* = 12.23$, on the left (ISONylon10) and a $Q_d^* = 12.74$, $\phi = 7.53$, and $H/D^* = 2.17$, on the right (ISOPropD14). 113

Figure 6.1: Comparison of measured heat release rate for three different liquid fuels, Heptane, Iso-Propanol, and Toluene. 116

Figure 6.2: Plot of heat of combustion of each fuel verses measured and ideal heat release rate for three liquid fuels. 116

Figure 6.3: Comparison of measured front (solid lines) and rear (dashed lines) temperatures for three different liquid fuels, Heptane, Iso-Propanol, and Toluene. 117

Figure 6.4: Comparison of measured front (solid lines) and rear (dashed lines) oxygen volume fraction as a function of time for three different liquid fuels, Heptane, Iso-Propanol, and Toluene. 117

Figure 6.5: Comparison of measured front (solid lines) and rear (dashed lines) CO volume fraction as a function of time for three different liquid fuels, Heptane, Iso-Propanol, and Toluene. 118

Figure 6.6: Comparison of measured front (solid lines) and rear (dashed lines) CO_2 volume fraction as a function of time for three different liquid fuels, Heptane, Iso-Propanol, and Toluene. 118

Figure 6.7: Comparison of measured front (solid lines) and rear (dashed lines) total hydrocarbon (THC, on a CH_4 basis) volume fraction as a function of time for three different liquid fuels, Heptane, Iso-Propanol, and Toluene. (Note: Front Iso-propanol results are not shown as the analyzer failed during this test. 119

Figure 6.8: Comparison of the measured radiative heat flux at the front ceiling location, channel ID:HFFCE, for heptanes, toluene, and isopropanol................................ 119

Figure 6.9: Comparison of the measured radiative heat flux at the front floor location, channel ID:HFFFL, for heptanes, toluene, and isopropanol. 120

Figure 6.10: Comparison of the measured heat release rate for polystyrene and poly propylene solid fules. 122

Figure 6.11: Comparison of measured front (solid lines) and rear (dashed lines) temperature as a function of time for two different solid fuels, polystyrene and polypropylene. 122

Figure 6.12: Comparison of measured front (solid lines) and rear (dashed lines) oxygen volume fraction as a function of time for two different solid fuels, polystyrene and polypropylene. 123

Figure 6.13: Comparison of measured front (solid lines) and rear (dashed lines) CO_2 volume fraction as a function of time for two different solid fuels, polystyrene and polypropylene. 123

Figure 6.14: Comparison of measured front (solid lines) and rear (dashed lines) CO volume fraction as a function of time for two different solid fuels, polystyrene and polypropylene. 124

Figure 6.15: Comparison of measured front (solid lines) and rear (dashed lines) total hydrocarbon volume fraction (on a CH_4 basis) as a function of time for two different solid fuels, polystyrene and polypropylene. 124

Figure 6.16: Comparison of heat release rate for heptane fires with a single burner and two distributed burners. 127

Figure 6.17: Comparison of front (solid line) and rear (dashed line) temperatures for heptane fires with a single burner and with two distributed burners. 127

Figure 6.18: Comparison of front (solid line) and rear (dashed line) O_2 volume fraction for heptane fires with a single burner and with two distributed burners. 128

Figure 6.19: Comparison of front (solid line) and rear (dashed line) CO_2 volume fraction for heptane fires with a single burner and with two distributed burners. 128

Figure 6.20: Comparison of front (solid line) and rear (dashed line) CO volume fractions for heptane fires with a single burner and with two distributed burners. 129

Figure 6.21: Comparison of front (solid line) and rear (dashed line) total hydrocarbon volume fraction for heptane fires with a single burner and with two distributed burners. 129

Figure 6.22: Comparison of front floor (solid line) and center ceiling (dashed line) heat fluxes for heptane fires with a single burner and with two distributed burners. 130

Figure 6.23: Comparison of heat release rates for three different ventilation conditions of a heptane spray burner fire operating at a nearly steady state of 1000 kW 132

Figure 6.24: Comparison of front (solid likes) and rear (dashed lines) temperatures for three different ventilation conditions of a heptane spray burner fire operating at a nearly steady state of 1000 kW. 132

Figure 6.25: Comparison of front (solid likes) and rear (dashed lines) O_2 volume fraction for three different ventilation conditions of a heptane spray burner fire operating at a nearly steady state of 1000 kW. 133

Figure 6.26: Comparison of front (solid likes) and rear (dashed lines) CO_2 volume fraction for three different ventilation conditions of a heptane spray burner fire operating at a nearly steady state of 1000 kW. 133

Figure 6.27: Comparison of front (solid likes) and rear (dashed lines) CO volume fraction for three different ventilation conditions of a heptane spray burner fire operating at a nearly steady state of 1000 kW. 134

Figure 6.28: Comparison of front (solid likes) and rear (dashed lines) total hydrocarbon volume fraction for three different ventilation conditions of a heptane spray burner fire operating at a nearly steady state of 1000 kW 134

Figure 6.29: Comparison of ideal heat release rate, determined from fuel flow rate, and measured heat release rate measured from oxygen loss caloremetry for test ISOHept27, where the heat release rate was ramped linearly with time 136

Figure 6.30: Comparison of temperature and various species mole fractions for ISOHept27 test with a linear heat release rate ramp. 136

Figure 6.31: Comparison of various gas species from both gas analyzers and GC analysis with heat release rate for heat release rate ram test ISOHept27 137

LIST OF TABLES

Table 2.1: Total delay times for the three gas sample probes used in the experiment. All delay times have an expanded uncertainty of ±2 s. 15

Table 2.2. List of micro-GC columns, specifications, and GC parameters used during FSE experiments. 17

Table 2.3. List of calibration standards and precision analyzed gases that were utilized for GC calibration. 17

Table 2.4: Sequence of controls for storing and then analyzing gas samples in the gas sample storage system. 21

Table 2.5. Location of measurement probes inside of the enclosure. 32

Table 2.6: Uncertainty of measurements 35

Table 3.1: List of test conditions considered in this report. 38

Table 3.2: Description of calorimetry measurement labels. 40

Table 3.3: Summary of averaged steady-state results of HRR and exhaust stack species measurements. U indicates the standard deviation in each steady state measurement. 49

Table 3.4. Description of interior gas temperature measurement labels. 52

Table 3.5: Summary of averaged steady-state results of temperatures at front locations. ... 58

Table 3.6: Summary of averaged steady-state results of temperatures at rear locations. 59

Table 4.1. Stoichiometric Value of the Mixture Fraction (Zst) for different fuels. 81

Table 4.2: Average fractional soot, CO and CO/soot ratio at the front and rear compartment measurement locations. 99

Table 4.3: Average yields of soot, CO and CO/soot ratio at the front and rear compartment measurement locations. 101

Table 4.4: Summary of averaged steady-state results of combustion efficiency in the exhaust stack. The steady state periods here are the same used for all steady state measurements and are listed in Table 3.5. The uncertainty, U, indicated here only reflects the statistical variation 109

Table 5.1: List of Normalized Scaling Parameters for Compartment Fires and the Range of Values Examined in this Study. 111

Table 5.2: Scaling comparison between the reduced scale enclosure (RSE) and the full scale enclosure (FSE) for heptane fires. The heat release rate values (Q) are taken as the heat release measured when CO began to be measured in the room. 113

Table 6.1: Heat of combustion and measured and ideal heat release rates for liquid fuels. 115

Table 6.2: Fuel Properties (r is the stoichiometric ratio of fuel to air). 125

1 INTRODUCTION

This report describes new full-scale compartment fire experiments, which include local measurements of temperature, heat flux and species composition, and global measurements of heat release rate and mass burning rate. The measurements are unique to the compartment fire literature. By design, the experiments provided a comprehensive and quantitative assessment of major and minor carbonaceous gaseous species and soot at two locations in the upper layer of fire in a full scale ISO 9705 room [1].

Fire protection engineers, fire researchers, regulatory authorities, fire service and law enforcement personnel use fire models (such as the NIST Fire Dynamics Simulator, FDS[2]) for design and analysis of fire safety features in buildings and for post-fire reconstruction and forensic applications. Fire field models have historically showed limited ability to accurately and reliably predict the thermal conditions and chemical species in underventilated compartment fires. Formal validation efforts have shown that for well ventilated compartment fires, with the exception perhaps of soot, field models do quite well in predicting temperature and species when experimental uncertainty is accounted for. Inaccurate predictions of incomplete burning and soot levels impact calculations of radiative heat transfer, burning rates, and estimates of human tenability. High-quality (relatively low, quantified uncertainty) measurements of fire gas species, temperature, and soot from the interior of underventilated compartment fires are needed to guide the development and validation of improved fire field models.

The experimental results provided in this report are the continuation of a long-term National Institute of Standards and Technology (NIST) project to generate the data necessary to test our understanding of fire phenomena in enclosures and to guide the development and validation of field models by providing high quality experimental data. The experimental plan was designed in cooperation with developers of the NIST FDS model to assure that the measurements would be of maximum value. Advanced development of FDS and other field models is extremely important, since it will lead to improved accuracy in the prediction of underventilated burning, typical of fire conditions that occur in structures. Improving models for under-ventilated burning will foster improved prediction of important life safety and fire dynamic phenomena, including fire spread, backdraft, flashover, and egress (involving the presence of toxic gas and smoke), which are critically important for application of fire models for fire safety.

1.1 Motivation and Objective

Field models, such as the NIST Fire Dynamics Simulator (FDS)[2] are widely used by fire protection engineers to predict fire growth and smoke transport for practical engineering applications. Many field models numerically solve the conservation equations of mass, momentum and energy that govern low-speed, thermally-driven flows with an emphasis on smoke and heat transport from fires. Among the various assumptions used in the development of early versions of FDS, all chemical species were tied to a single mixture fraction variable by use of a set of mixture fraction state relations. A single mixture fraction variable cannot be used for the prediction of carbon monoxide and soot, and the yield of these species was prescribed in FDS 4, rather than predicted. In fact, the yield of these species is usually not constant, but a complex function of their time-temperature history. In practice, an knowledgeable user would attempt to pick yields that would reflect the anticipated ventilation condition of the simulation from literature values for well-ventilated burning, using data from a bench-scale apparatus [3] or from other sources such as the full scale experimental results presented here. Using this approach, the CO volume fraction for pool fire burning in an under-ventilated compartment can be underestimated by as much as a factor of ten.

FDS 5 [2] has included a simple predictive method for CO production. This revised method breaks the mixture fraction calculation into two parts resulting in a two-step chemistry model. This change in the chemistry of the model is an improvement over the prescriptive method used in FDS 4, however still under predicts CO substantially. A recent paper [4] by the developers of FDS reported on the model validation of the reduced scale enclosure (RSE) experimental results [5]. They found that FDS 5 has improved its prediction of fires in this configuration. The best agreement was observed with methanol, a very low sooting fuel. In general good agreement was observed with velocity and temperature data from these experiments. The CO production model was improved substantially, however there is still significant difference between the experiments and the model. As more soot was produced by the fuels and the fires became more underventilated an under prediction of CO and an over prediction of CO_2 was observed. The authors attributed these effects to the specific assumptions being made in the FDS CO prediction scheme.

In an effort to validate current fire models and to further the development of better predictive methods for fires, the current report presents new and unique data on full scale underventilated compartment fire experiments which builds on the previous data concerning reduced scale enclosures (RSE) [5]. The experiments are presented with analysis and experimental modeling results as a method of explaining the fire behavior and aiding in analysis.

1.2 Previous Work

Experimental research on enclosure fires has been on-going in fire research laboratories and academic institutions over the last 50 years. The motivation has varied from applied investigations studying particular fire scenarios to more fundamental work with the goal of understanding toxic species production behavior in fires. Some of the fundamental research that tried to ascertain ventilation and upper-layer effects on enclosure fire chemistry was conducted in well-controlled hoods. Sometimes, the main objectives of this research was to generally develop and validate fire models or particular structural fire simulations, while much of the research was conducted to acquire a better understanding of complex enclosure fire dynamics with a focus on chemical and thermal conditions. This section provides an overview of some of the recent research efforts in enclosure fires and highlights some of the more pertinent experimental work.

Research conducted at Harvard University and the California Institute of Technology in the 1980s explored fires burning under an exhaust hood (false ceiling) to simulate the layer effect of an enclosure fire, e.g. [6, 7]. The relative distance of the fire below the hood was adjusted to vary the entrainment of air into the plume before it entered the upper layer. These experiments focused on underventilated burning, pathways for air to enter the upper layer, and the validity of the concept of "global equivalence ratio" (GER) which is the fuel-to-air mass ratio normalized by the mass ratio required for stoichiometric burning. Some recent modeling work by Cleary and Kent [8], has also focused on experimental data from hoods. In a recent study, Brohez et al.

explored the use of a bench-scale calorimeter to measure fire properties of materials burning in underventilated conditions [8, 9].

Research at NIST by Bryner et al. further explored the global equivalence ratio concept and carbon monoxide production in a reduced (2/5) scale enclosure with natural gas as the principal fuel [9]. The results showed that the upper layer in enclosure fires is not homogeneous, and that CO can be produced in greater quantities than predicted by the GER concept, depending on temperatures and flow patterns developed within an enclosure. The previous effort [5] was meant to overlap some of the conditions explored by Bryner et al. and to repeat and fill gaps in the data. Pitts expanded the work to full-scale and other fuels such as heptane and wood. It was established that wood pyrolysis in the upper layer of an enclosure fire can produce high concentrations of CO directly without further oxidation to CO_2 [10]. A subsequent study by Lattimer confirmed and expanded on this research [11].

Researchers at Virginia Tech investigated fires in a reduced-scale enclosure that directed the air inflow through slots in the floor connected to a duct where instrumentation was used to quantify air entrainment [12]. Several fuels were studied, and this configuration produced results consistent with GER predictions due to the more distinct, less dynamic nature of the gas layer structure. Later work used a more typical enclosure design and focused on transport of gas species outside the doorway and how it was affected by doorway geometry, soffit design, and hallway configuration [13]. More recently, Gann et al [14] conducted research on transport of toxic species in a full-scale enclosure with a corridor. These data were analyzed by Hirschler [15]. Researchers in Sweden conducted a study [16] of under-ventilated fires in an ISO 9705 room with a window vent of varying height. Several polymer fuel types were included in this study and measurements of local equivalence ratio and toxic gas species were performed.

Pitts [17] provides a comprehensive review of the application of the GER concept to predict CO concentration in building fires, using data from the Harvard and Cal Tech hood experiments [3, 9], the Virginia Tech enclosure studies [11], and the NIST reduced-scale enclosure experiments [10, 19]. Several CO formation mechanisms were identified, which were substantiated by detailed chemical kinetic modeling. While the GER concept is of limited utility for predicting the local CO concentration, important aspects of enclosure fire dynamics and chemistry are highlighted in this paper.

Several recent experimental studies [18-20] have used very small scale enclosures (0.21 m^3, 0.06 m^3, and 0.05 m^3, respectively) while investigating under-ventilated burning of propane and heptane fires. These bench-scale studies described the structure and dynamics of under-ventilated burning including extinction, flame projection and flame stability. Another recent study [21, 22] has used an intermediate-scale enclosure similar to that used for this paper, but a roof vent was added as well.

Most recently a previous component of this research project focused on similar experimental measurements of a Reduced Scale Enclosure (RSE[1])[5]. The RSE was a 2/5 scale ISO 9705 room designed based on the previous studies of Bryner [9]. Similar to Bryner's experiments, natural gas served as a fuel; the burning of heptane, toluene, methanol, ethanol, and polystyrene

---

[1] The data from this set of experiments is currently available online at http://www.fire.nist.gov/testdata/RSE/

was also investigated. In most experiments, the fuel was controlled and metered by flow valves or pumped into a pool burner or spray nozzle. Experiments were run to near-steady conditions. Multiple fire sizes were run consecutively to decrease the time required to approach steady-state. Ventilation was varied during some experiments by modifying the door opening. Two types of enclosure lining materials were investigated and compared.

Recently, NIST has conducted a number of high-profile case studies in which realistic-scale mock-ups of actual fire scenarios were recreated with the ultimate goal of improving building codes and standards. These studies included the World Trade Center disaster investigation [23], the Rhode Island Station nightclub fire [24], and the Chicago Cook County Administration Building fire investigation [25]. The compartment fires in all of these studies burned real furnishings and became under-ventilated as the fire evolved. In addition, a series of large-scale compartment fire experiments were conducted to simulate an over-ventilated fire in a nuclear power plant cable room [26] to provide data for fire model validation.

1.3 Experimental Scope

While some previous studies have considered the mixture fraction to analyze experimental compartment fire data, few have considered minor hydrocarbon species and with the exception of Ref. [5] none have considered soot. In tandem, accurate measurements of temperature at these same locations allowed analysis of thermal effects on species concentrations. A wide range of fuel types were considered, including aliphatic hydrocarbons (natural gas and heptane), aromatic hydrocarbons (toluene, polystyrene) and an alcohol (isopropanol).

The series of experiments reported on here was conducted in a full scale (ISO 9705 room) enclosure (FSE). The enclosure defined in the international standard ISO 9705 "Full-scale room test for surface products" [1] is an important structure in which to conduct fire research. The experiments repeated and extended a part of the work of Bryner and coworkers [9] as well as the authors previous work with a reduced scale enclosure [5]. Similar to Bryner's experiments, natural gas served as a fuel; the burning of heptane, toluene, iso-propanol, polypropylene, nylon, and polystyrene were also investigated. The fuel was either allowed to burn freely in a pan or controlled and metered by flow valves or pumped into a pool burner or spray nozzle. Experiments were either run as free burns or at near-steady conditions. Multiple fire sizes were run consecutively to decrease the time required to approach steady-state. Ventilation was varied during some experiments by modifying the door opening.

Temperature and species composition measurements in the current experiment were made at many of the same nominal locations (scaled where appropriate) as studied previously by Bryner et al,[9] and Bundy et al [5]. Measurements included O_2, CO, CO_2, total hydrocarbons, temperature, and heat fluxes. One emphasis of this series was to further develop techniques for the measurement of hydrocarbons and soot. Hydrocarbons were measured with FID analyzers (total hydrocarbons) and gas chromatography (GC). The GC measurements were used to independently validate the total hydrocarbon measurements and to allow accurate determination of species mass distribution. The quantification of hydrocarbon species was needed to describe the chemical structure of under-ventilated fires. Soot samples were extracted from within the enclosure and measured gravimetrically. Optical soot measurements were also performed.

The fuels included in this test series were selected to cover a wide range of combustion properties and to simulate fuels encountered in actual building fires. Gases, liquids, and solids were selected for testing to cover a wide range of physical properties. Realistic materials represent complex multi-component fuels. In this study, all of the fuels selected were homogeneous single component fuels to simplify the analysis and to attempt to find generalizable trends in the results. Real materials are often oxygenated. This includes many types of commodity materials including nylon (e.g., carpet), cellulose (e.g. paper and building products), polyester (e.g., fabric), epoxy (e.g., adhesives), polymethylmethacrylate (PMMA), and polyoxometalate (POM). In this study, iso-propanol was selected as a surrogate to represent the compartment fire chemistry in the burning of oxygenated fuels.

In a real compartment fire, fuel sources are physically distributed throughout the compartment. In this study, a multiple locations for the fuel were used to simulate this effect for the purpose of model validation. Multiple fuel locations allow for comparison with single fuel sources and add value to the overall research product.

Heat feedback and natural ventilation give rise to important aspects of the structure and dynamics of the fire, such as the temperature field and the spatial distribution of combustion products. This study deliberately set out to investigate representative fire conditions at two key locations in the upper layer of the compartment, which were selected based on the geometrically scaled locations used previously [5]. The upper layer locations were selected to provide two distinct conditions in the upper layer, one relatively close to the natural ventilation flow of fresh air through the doorway and the other relatively far from the doorway, on the far side of the fire source. Combined gas species and temperature measurement probe was situated in the room and moved along a vertical line to provide additional data. The vertically moving probe allowed for sampling in the upper layer, lower layer or within the transition as deemed useful by the researchers. Two thermocouple trees were also situated in the room in order to evaluate the vertical thermal profile within the room. To enhance the range of conditions investigated and in an attempt to seek information on the relationship between the combustion products and local fire conditions, a broad range of fire heat release rates and a number of very different fuel types were selected for study. At the same time, the effect of compartment ventilation was changed to induce a range of mixing and compartment fire conditions.

2 EXPERIMENTAL DESIGN

2.1 Design of the room

The dimensions of the ISO 9705 room were used in this experiment due to its wide utilization in other works and to build upon the previous experiments with the reduced scale enclosure (RSE) which was purposefully scaled to 2/5 of the ISO 9705 dimensions. The RSE investigation looked at a variety of room construction materials and helped to guide the development of the final design of the ISO 9705 room which was used here.

2.1.1 Dimensions

The experiments discussed in this report are based on the dimensions of the ISO 9705 room [1]. The full scale enclosure (FSE) is illustrated in Figure 2.1. The design internal dimensions of the room were set to the ISO 9705 standard to be 240 cm x 240 cm x 360 cm with a doorway of 80 cm x 200 cm. The floor of the enclosure was raised 35 cm above the ground. Fractional doorways were also utilized with a Doorway Fraction, DF, defined as the fractional width (and therefore area) of the 80 cm door. The height of the door was not varied. Due to the nature of the lining material and the fasteners used to hold it in place there is some variability in the actual dimensions of the room. However, the as built dimensions were measured extensively and all uncertainty was found to be within ± 2 cm, well within the tolerance of the ISO 9705 standard of ± 5 cm. Additional measurements were taken periodically within the room during the experimental tests and never exceeded an uncertainty of ± 2 cm.

2.1.2 Materials

The support structure of the room was built using 20 gauge (0.89 mm) steel structural studding and 20 gauge (0.89 mm) sheet steel. The floor of the structure was constructed of 0.48 cm thick steel sheet metal. The actual room used in the experiments can be seen in Figure 2.2. The studs and sheet metal were built such that their internal dimensions were 10 cm greater than the ISO 9705 standard. On top of the sheet metal (on the interior surfaces) were installed two layers of 2.5 cm thick, 128 kg/m^3 density, ceramic fiber blanket, K-lite^{tm} HTZ. The blanket was composed of 30 % AL_2O_3, 54 % SiO_2, 16 % ZrO_2, and trace amounts of other components. The uncertainty in the composition of the primary four components is ± 1 %. The ceramic fiber blanket was held in place by alumina ceramic (99 % Al_2O_3) insulation retainers (Refractory Anchors Inc. model RA38) with a depth of 5 cm. These anchors are shown in Figure 2.3. The ceramic anchors were secured to the sheet steel wall with self-tapping sheet metal screws and washers. Insulation retainers were installed in the ceiling studs, spaced 40.5 cm, at 30.5 cm intervals along each stud. On the walls the insulation retainers were also installed with an arrangement of 40.5 cm by 30.5 cm near the top of the wall with the spacing increasing to 40.5 cm by 70 cm as the retainer placement approached the floor. Extra retainers were placed as necessary to hold edges and corners securely in place.

This structure design proved to be quite robust. Through a series of 24 tests only minor repairs to the blanket and ceramic retainers were necessary. The steel skin and steel studs held up well with the exception of the portions of the structure framing the doorway. Figure 2.2 shows that in the vicinity of door way the ceramic fiber insulation was wrapped around the doorway to protect it from the heat and radiation from the room. This additional insulation was not sufficient to protect the studs from excessive heat causing them to soften and deform over time. This situation was further exacerbated by the convective heat transfer from the hot, fast moving gasses leaving the enclosure.

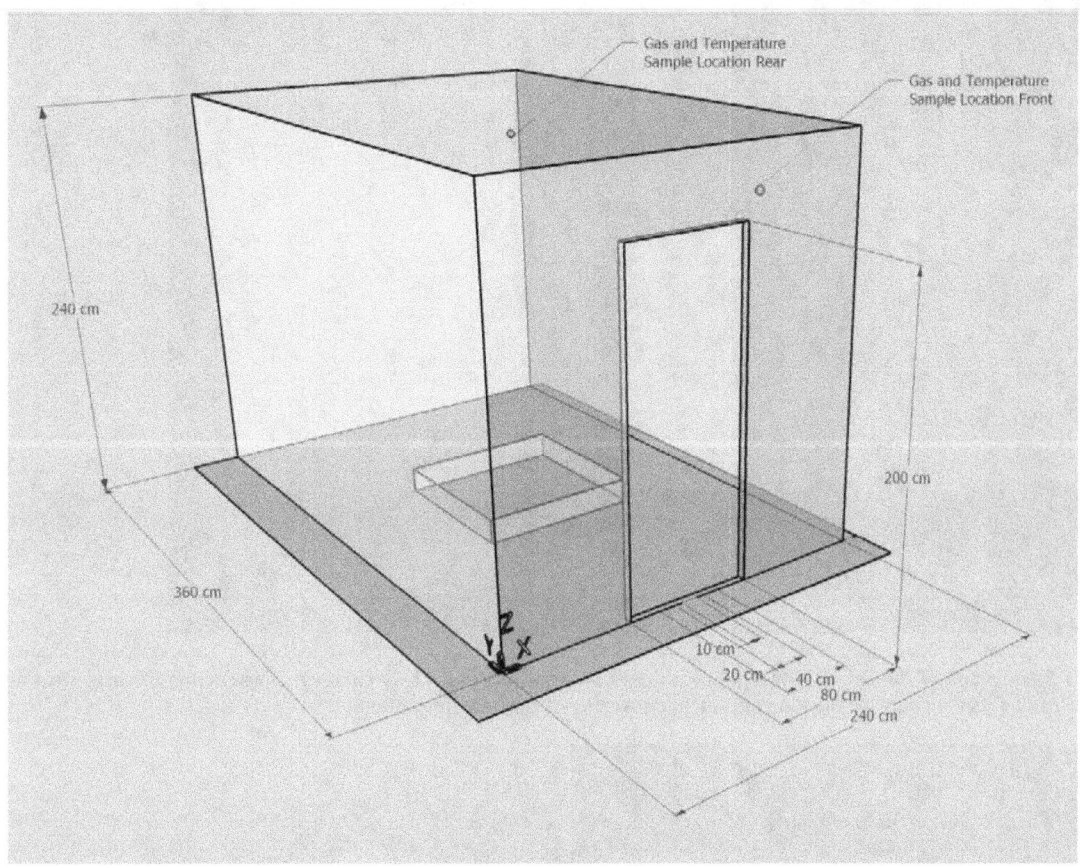

Figure 2.1 Internal dimensions of ISO 9705 enclosure used in these experiments including multiple door widths and gas sample and temperature probe locations. All dimensions have an uncertainty of ± 2 cm.

Figure 2.2 Photograph of the actual ISO 9705 room used for experiments. The structural construction of sheet steal on steel studs can be seen along with the internal surface covering of ceramic fiber blanket.

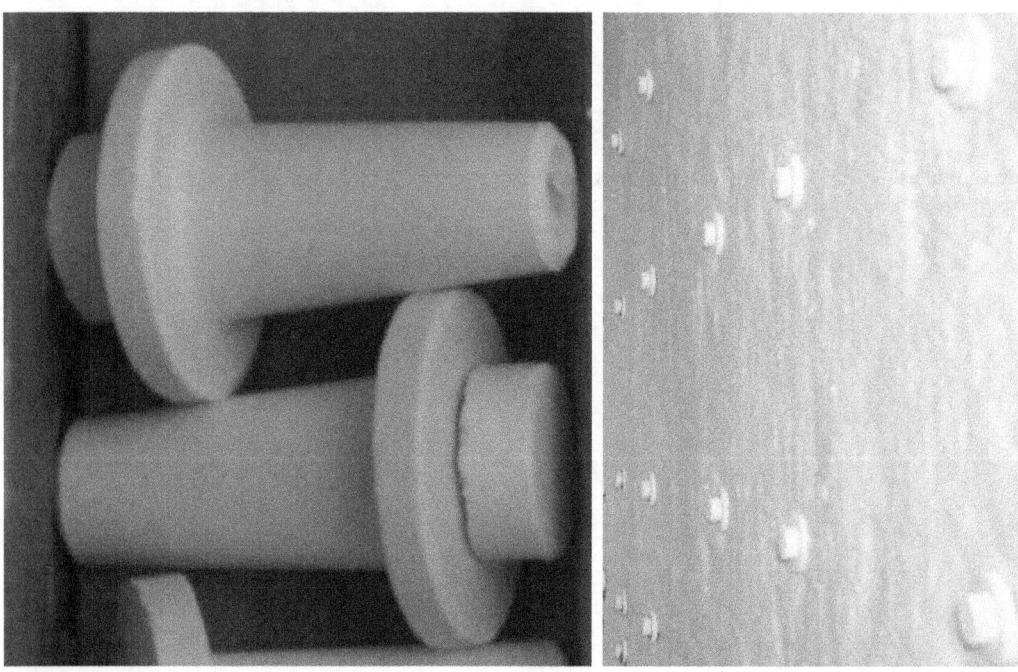

Figure 2.3 Ceramic insulation retainers used to secure the ceramic fiber blanket to the sheet steel walls. The actual retainer is shown (left) as well as its installed configuration (right).

2.1.3 Doorway Dimensions

A 20 cm doorway (1/4th of the ISO 9705 standard) was used for the majority of the experiments in order to force the room to reach under-ventilated conditions with a smaller fire size and therefore limiting the temperatures and thermal radiation within the room. Several other doorway widths were also utilized in order to evaluate the effect of the vent area. In addition to the 1/4 width (20 cm) door, a full ISO 9705 door (80 cm) width, a 1/2 width (40 cm), and a 1/8 width (10 cm) doorway were considered. The height of each doorway was held constant at 200 cm. The ISO 9705 structure was not modified in order to create the different door widths, instead inserts were constructed from steel studding and ceramic fiber blanket. The doorway inserts allowed for a quick change of the door size as well as keeping the doorway accessible for work inside the room. Unfortunately, repeated heating and cooling of the door inserts resulted in deformations. Every attempt was made to ensure that proper door sizing was maintained, however due to variations there is an uncertainty of ± 10 % in the doorway widths.

The specific door dimensions for each test are included in the description of test conditions (refer to Table 3.1). In some places in the document the doorway fraction (DF) was used to indicate the width of the doorway utilized.

2.1.4 The Burners

Two primary types of burners were utilized, free-burn fuel pans and a spray burner. Additionally, a pump-fed, water cooled liquid burner and a gravel filled natural gas burner were also utilized for a limited number of tests.

The free-burn pan type burners were constructed from welded sheet steel 0.635 cm thick. Two each of burners with internal dimensions sized at 50 cm x 50 cm (0.25 m^2), 70.7 cm x 70.7 cm (0.5 m^2), and 100 cm x 100 cm (1 m^2) each with a 10 cm lip were constructed as illustrated in Figure 2.4. The uncertainty in the burner dimensions was ± 0.5 cm, not taking into account warping that occurred during the experiments. The burners were used individually and in matched pairs to simulate single and distributed fuel loads. The burners were positioned in the geometric center of the floor (position 1) and/or along the centerline of the room next to the rear wall (position 2) as illustrated in Figure 2.5. The single burner cases were also moved between the two discrete positions (1 and 2) to simulate various fuel package locations.

Each of the free-burn fuel pans was mounted through the floor to one of two load cells, Mettler Toledo, Jaguar KCC150 or KCC300, each with a measurement accuracy of ± 1 g. In this way the fuel loss rate could be measured to calculate the ideal heat release rate and combustion efficiency of the free burn cases. Additionally this configuration allowed for a measurement of fuel being collected in the pan beneath the spray burner for those cases.

One 0.5 m^2 (70.7 cm x 70.7 cm) pan in position 1 (cf. Figure 2.5) was also used with the spray burner configuration as well. Different spray nozzles were utilized depending on the desired fuel flow rate. All spray nozzles were BETE Low Flow/Full Cone Whirl nozzles. WL 1 and WL 1-1/2 were both utilized, both were constructed from stainless steel and featured a 90 degree cone angle. The pump flow rate was varied in order to provide different flow rates at the nozzle to produce different fire sizes. A load cell was utilized in the spray burner configuration in addition to the pump flow-rate monitoring to determine if any fuel collected in the pan. In this way all of the fuel from the spray burner could be accounted for and measured to calculate the overall combustion efficiency.

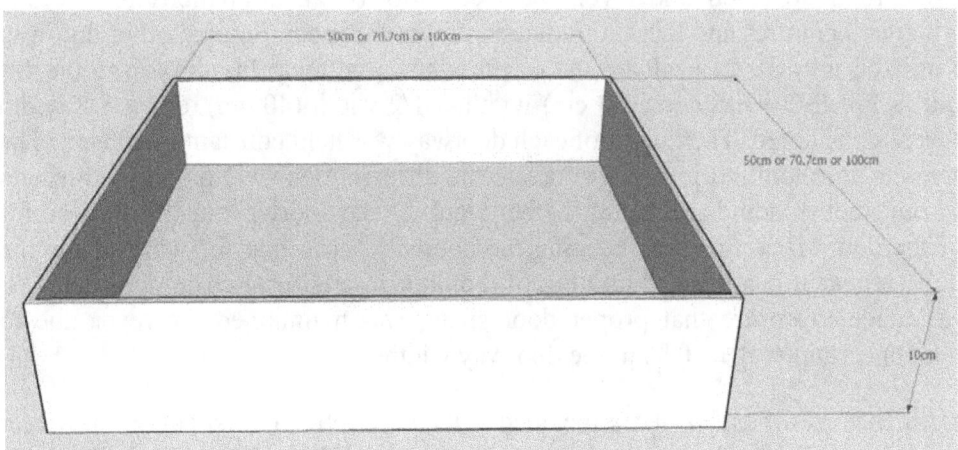

Figure 2.4: Free-burn and spray burner pan construction and dimensions. The dimensions of 50 cm, 70.7 cm, and 100 cm are all internal burner dimensions. All burners had a lip height of 10 cm.

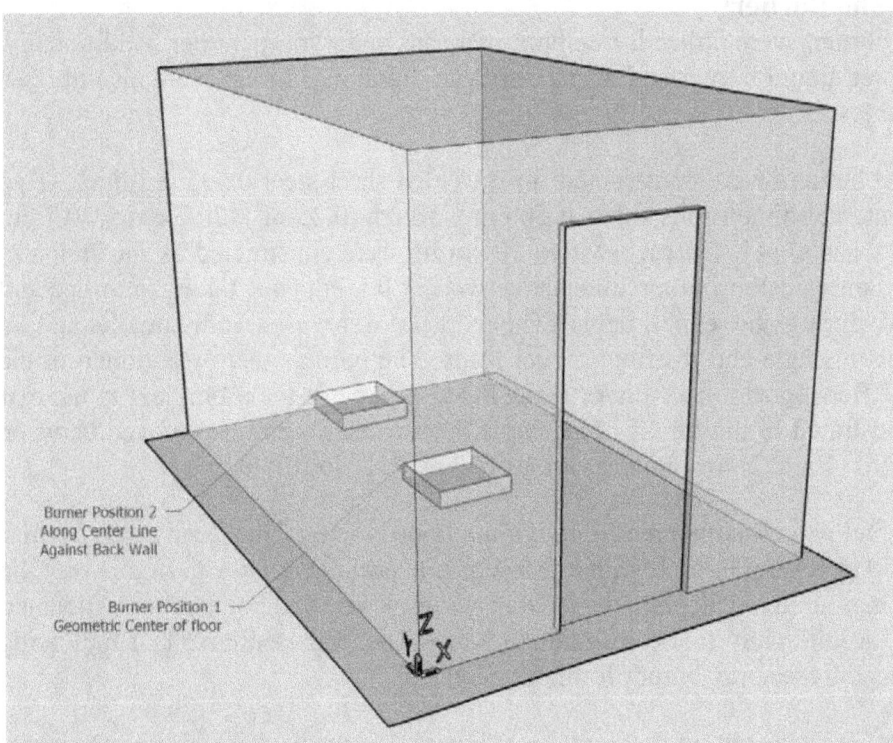

Figure 2.5: Positioning of free-burn pan burners. The burners of different size were placed at the geometric center of the floor and/or along the centerline against the back wall. Pans in either position were mounted on load cells to measure the mass loss (or gain in the case of the spray burner) to determine fuel loss rate. For the spray burner cases only a single 70.7 cm x 70.7 cm pan was used in the center of the floor to catch the fuel spray. Both the pump fed pool burner and the gravel filled natural gas burner were located at position 1 inside the room.

2.2 Overview of equipment

2.2.1 Heat Release Rate Measurement

Heat Release Rate (HRR) measurements were conducted using the 6 m × 6 m calorimeter at the NIST Large Fire Research Laboratory (LFRL). The HRR measurement was based on the oxygen consumption calorimetry principle first proposed by Huggett [27]. This method assumes that a known amount of heat is released for each gram of oxygen consumed by a fire. The measurement of exhaust flow velocity and gas volume fractions (O_2, CO_2 and CO) were used to determine the HRR based on the formulation derived by Parker [28]. A detailed description of the methodology used for this measurement can be found in a previous report [29]. In 2001, the 6 m × 6 m square hood was installed in the LFRL. A schematic drawing of the 6 m square hood is shown in Figure 2.6. The exhaust flow rate and extractive gas measurements were performed in a horizontal straight section of the 152 cm diameter duct on the roof of the large fire lab. Six bi-directional probes, located on the vertical centerline, were used to measure the exhaust flow velocity. Because of the non-uniform shape of the velocity profile, a flow calibration coefficient was used in the HRR calculation. The flow coefficient was determined using a natural gas burner to conduct a five point calibration before and after the test series. The flow calibration coefficients ± 2σ for these tests ranged from 0.906 ± 0.04 to 0.933 ± 0.05. The calibrations were performed over a range of fire sizes from 500 kW to 3000 kW. The exhaust mass flow rate for the experiments described here varied from 12 kg/s to 17 kg/s.

Exhaust gases was sampled through a perforated tube cross in a horizontal section of the duct downstream of the velocity probes. Figure 2.7 shows the exhaust gas sampling system. The main difference between this system and the one previously reported [29] was the method for removing water from the gas sample. The current system uses a Nafion® dryer instead of a dry ice cold trap. Nafion is a copolymer of tetrafluoroethylene (Teflon®) and perfluoro-3,6-dioxa-4-methyl-7-octene-sulfonic acid. A dew point meter was added to monitor the efficiency of the gas dryer. The dew point temperature meter measures the change in electrical impedance of a hygroscopic conductive polymer in the range of -80 °C to 20 °C. The delay time from the gas sample tube to the analyzers was 25 s. Measurements of exhaust soot and total hydrocarbons were not performed, because in most cases they have negligible effect on the HRR measurement. The combined expanded relative uncertainty of the HRR measurements reported here was 14 %, based on a propagation of uncertainty analysis [29]. The exhaust mass flow rate was the largest component of uncertainty in the HRR measurement. A list of commercial equipment used for all of the measurements described in this report can be found in Appendix B.

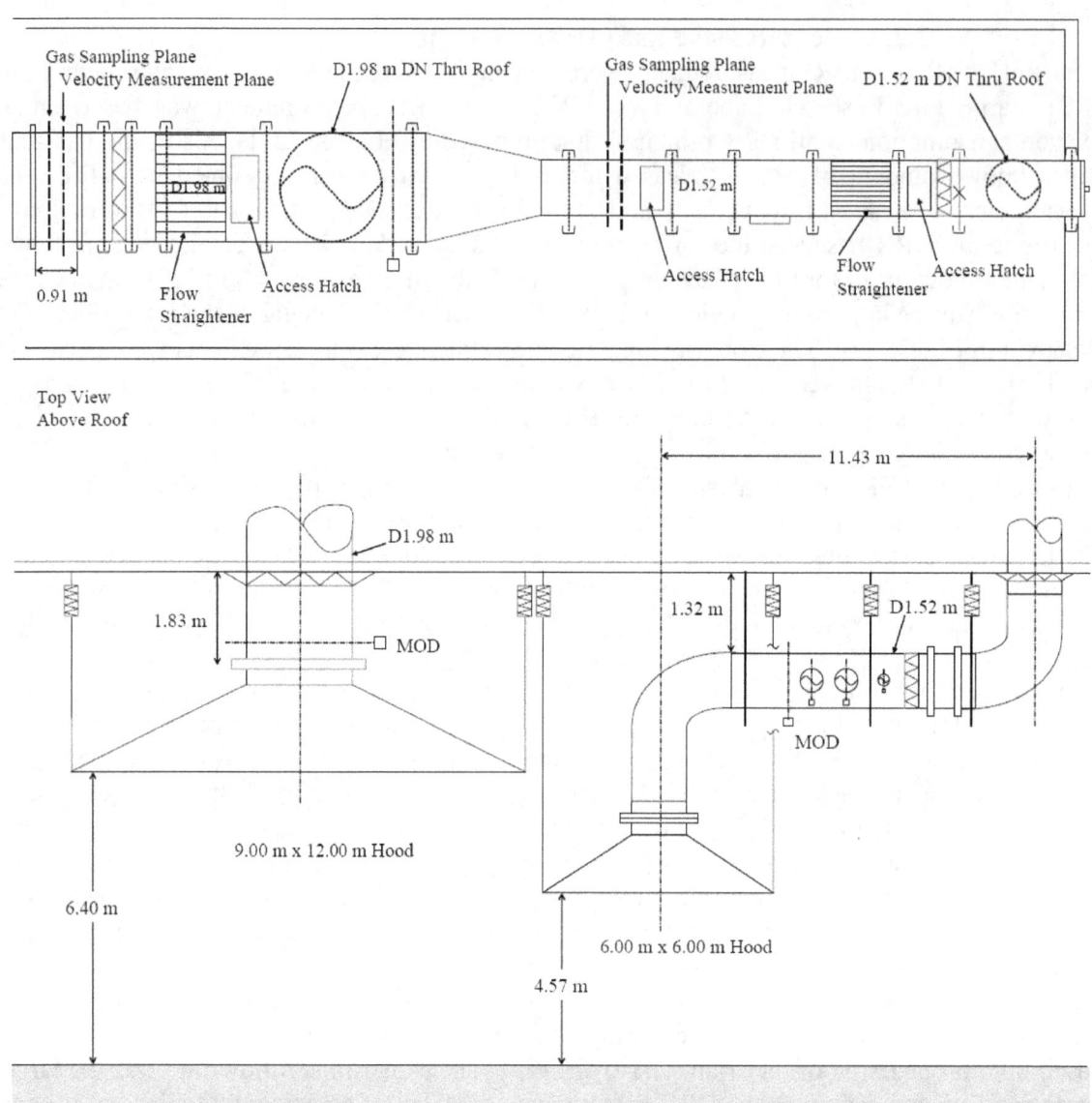

Figure 2.6: Schematic drawing of 6 m square hood and exhaust stack instrumented for calorimetry measurements. Taken from Ref. [29]

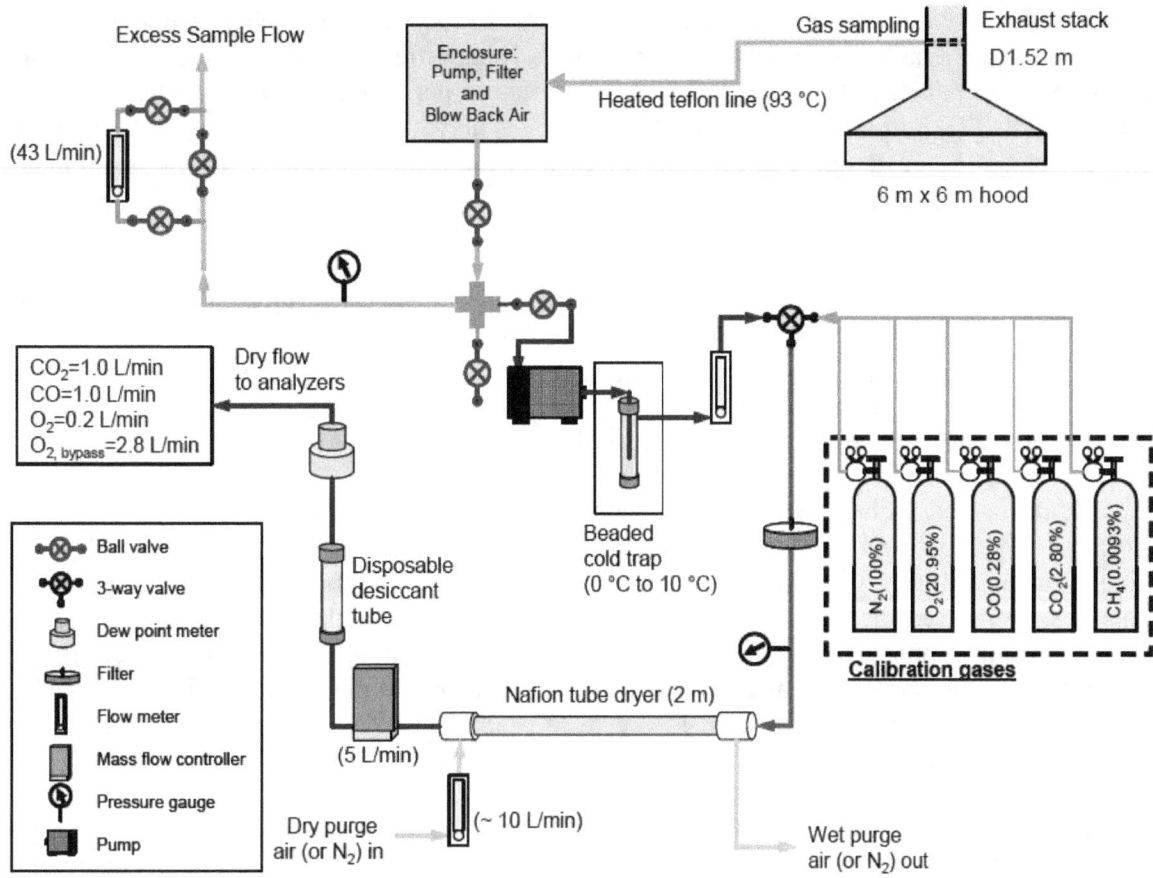

Figure 2.7: Exhaust gas sampling system used for heat release rate measurement.

2.2.2 Gas analyzers

Gas species were continuously measured at two locations (front and rear) inside the FSE during each of the tests and sometimes at a third location (vertically moving probe near rear on centerline). Oxygen was measured using paramagnetic analyzers. The 10 % to 90 % response time (t_{10-90}) of the oxygen analyzer was less than 12 s. Carbon monoxide and carbon dioxide were measured using non-dispersive infrared (NDIR) analyzers. The t_{10-90} response time for the CO_2/CO analyzers was less than 5 s. Total hydrocarbons were measured using two flame ionization detectors (FID) having a t_{10-90} response time of less than 1 s. A gas chromatograph (GC) was used intermittently during some of the tests at the front and rear gas sampling location. The cycle time on the GC measurements was 2 min. The dried sample gas dew point temperature was measured using a thin polymer sensor. Soot and temperature were also measured at these two locations. The total delay times for each of the analyzers were measured by initiating a small flame at the gas sample probe inlet and timing how long until a response was recorded by the gas analyzers. A summary of the delay times for each of the three sample probes discussed above is presented in Table 2.1.

The three total hydrocarbon analyzers used in these experiments were designed to measure high volume fractions of hydrocarbons. The analyzers were factory calibrated for up to 50 % volume fraction of hydrocarbons as methane and were capable of measuring even higher concentrations. The primary span gas used for these tests was 20 % volume fractions of methane with a balance of nitrogen. A span gas of 1 % methane was also used to periodically check the linearity of the detector. The FID burner fuel used was 40 % hydrogen and 60 % nitrogen on a volumetric basis. The expanded (k = 2) relative uncertainty of each of the span gas volume fractions, including CH_4, CO, CO_2, and O_2 was ± 1 %.

Each hydrocarbon analyzer had an internal filter to prevent soot from accumulating in the plumbing and internal sample pump which could lead to less sensitivity due to hydrocarbon contamination and also deterioration of some components of the instrument. It was later determined that additional external filtration of soot was necessary to protect the analyzer and enable a sufficient time period for sampling soot-laden flows. The external filter could be replaced much more frequently and easily than the internal filter.

Two liquid cooled probes were used to sample gas inside the enclosure at the front and rear locations. The 1 m long probes were constructed of 3 concentric stainless steel (type 304) tubes. Liquid coolant (water) was forced through the inner shell and returned through the outer shell. This design allowed the cooling fluid to condition the entire length of the probe. The inner diameter of the sample probe was 4.0 mm. The front and rear gas sample systems were identical, except the GC measurement was conducted continuously only at the front sample location and a gas sample storage system was used at the rear location. The moving sample probe did not include GC analysis.

The third, moving sample probe was constructed from an aspirated thermocouple. The aspirated thermocouple flow rate was set at 50 SLM and the gas sample was split off from that stream and pumped at 1 LPM to the gas analyzers.

The sample probes shown in Figure 2.8 were cooled using house water heated to 55 °C at a flow rate of 1 L/min. The total hydrocarbon analyzers were placed in the gas racks with the other analyzers. The gas sample stream water was removed with membrane dryers consisting of a bundle of Nafion tubes purged with dry air to selectively remove moisture from the sample stream. The Nafion conditioner has no effect on most of the gas species of interest, however, polar organic compounds (i.e. ketones and alcohols) are trapped by the dryer. A large area filter was added between the heated line and gas dryer to collect soot. Because the external filters and transfer lines after the gas dryer were not heated, there was a potential loss of high molecular weight hydrocarbons due to condensation. Due to limitations in the flow capacity of the dryer, the gas analyzers were connected in series. A mass flow controller set to 1 L/min was used to control the flow through the $O_2/CO_2/CO$ analyzers. The flow to the hydrocarbon analyzer was split prior to the mass flow controller. A 5 way ball valve was connected to each analyzer to switch between the gas sample, zero calibration gas and span calibration gas. A dew point transducer was connected to the sample gas line prior to the oxygen analyzer. The oxygen analyzer had separate inlet ports for zero and span gases. A needle valve was used to set the total flow to 3 L/min (only a small fraction of this passed through the FID). The bypass flow from the hydrocarbon analyzer was connected to the injection port of the GC (front sample location) or the gas sample storage system (rear sample location).

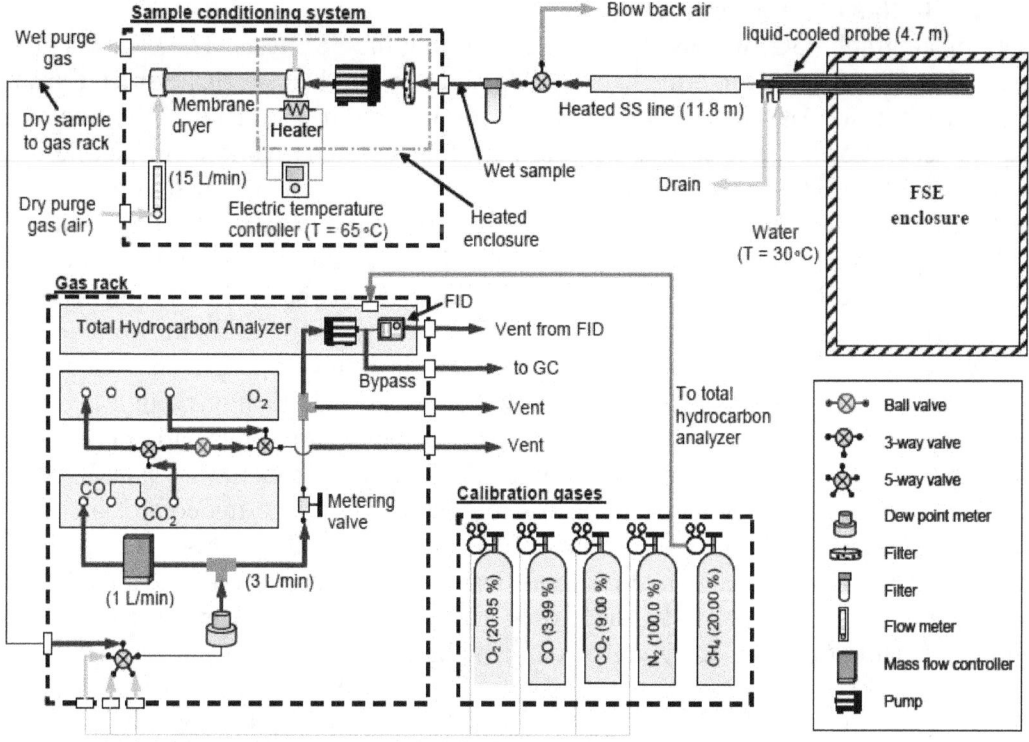

Figure 2.8: Schematic drawing of gas sampling system.

Table 2.1: Total delay times for the three gas sample probes used in the experiment. All delay times have an expanded uncertainty of ±2 s.

Channel	Delay time (s)
O2Rear	25
CO2Rear	21
CORear	21
UHRear	20
O2Front	23
CO2Front	19
COFront	18
UHFront	12
O2Move	19
CO2Move	15
COMove	14
UHMove	10

2.2.3 Gas Chromatography

A micro gas chromatograph (GC) was used periodically during the FSE tests. The GC was able to quantify a number of stable fuel, intermediate, and combustion product species extracted from the FSE during each test. An Agilent 300A micro-GC was used to quantify the species. Chromatographic separation of species was achieved by four columns working in parallel. A molecular sieve 5A, Plot-U, OV-1, and Stabilewax columns were used. Thermal conductivity detectors (TCD) were used on each column. A carrier gas of Helium was used on most of the columns while Argon was used on the mole-sieve column. Due to the similar thermal conductivity of Helium and Hydrogen it is difficult to distinguish between the two of them with the TCD, Argon, with a significantly different thermal conductivity, being different from any of the species we were expecting to find allowed for the quantification of a larger number of gas species. A summary of the different columns, their physical specifications, and the GC parameters used during analysis are listed in Table 2.2. Due to the nature of this particular GC, it allows for very fast methods to identify species, a method lasting only 2 minutes was utilized to provide a large number of data samples from each experimental run. The tradeoff to the high speed of the analysis was reduced sensitivity compared to conventional GCs. This means that the detail and separation of individual heavy hydrocarbons was sacrificed in order to provide a larger number of measurements of gas species such as hydrogen, methane, and nitrogen which are very important but not available from any other analysis utilized here.

Identification and quantification of gaseous species was accomplished by the use of gas phase calibration standards. Due to the wide variety of columns used and the large number of species and varying quantities that can be potentially identified, several different gas standards, listed in Table 2.3, as well as locally made calibrations for high concentrations of CO_2, H_2O, and CH_3OH were utilized to create the calibration for the GC.

Table 2.2. List of micro-GC columns, specifications, and GC parameters used during FSE experiments.

	Mole-Sieve	PlotU	OV-1	Stabilwax
Inlet	Backflush	Backflush	Variable	Variable
PreColumn	PlotU	PlotQ	None	None
Film	30.00 μm	10 μm	N/A	N/A
Diameter	320.33 μm	320.00 μm	N/A	N/A
Length	3 m	1 m	N/A	N/A
Film	12.00 μm	30.00 μm	2.00 μm	0.50 μm
Diameter	320.00 μm	320.00 μm	150.00 μm	250.00 μm
Length	10 m	8 m	8 m	10 m
Detector	TCD	TCD	TCD	TCD
Carrier Gas	Argon	Helium	Helium	Helium

Table 2.3. List of calibration standards and precision analyzed gases that were utilized for GC calibration.

Gas Standard Identification	Species	Quantity	Uncertainty
Restek Refinery Gas #2 SN: 480885	H_2, Ar, N_2, CO, CO_2, CH_4, C_2H_6, C_2H_4, C_2H_2, C_3H_8, → C_6H_{14}	0.10% -- 37.287%	±1 %
Scotty Lot No. 612201B Mix: 01-04-235--14	1-Butene, Ethylene, 1-Hexane, 1-Pentene, Propylene	1000 ppm	±5 %
Scotty Lot No. 612203B Mix: 01-04-224--14	N-Butane, Ethane, N-Hexane, Methane, N-Pentane, Propane	1000 ppm	±5 %

Analysis of variance (ANOVA) and regression analysis were employed in an effort to determine the uncertainty associated with the quantification of the gas species. The result of this analysis was the uncertainty of a single value, S_y, calculated from the calibration curve. This value was combined with the uncertainty of calibration gases and additional type B uncertainty and is presented as expanded uncertainty with a coverage factor of k=2 in the data set and the appropriate plots. The equations utilized for the analysis to determine S_y are show below:

$$S_y = \frac{S_{regression}}{\beta} \sqrt{\frac{1}{m} + \frac{1}{N} + \frac{(y_i - \bar{y})^2}{\beta^2 S_{xx}}} \tag{1}$$

where,

$$S_{regression} = \sqrt{\frac{(S_{yy}) - \left(\frac{(S_{xy})^2}{(S_{xx})}\right)}{N-2}} \tag{2}$$

$$S_{yy} = \sum y^2 + \frac{(\sum y)^2}{N} \tag{3}$$

$$S_{xy} = \sum xy + \frac{(\sum x \sum y)}{N} \tag{4}$$

$$S_{xx} = \sum x^2 + \frac{(\sum x)^2}{N} \tag{5}$$

m = number of measurements of the unknown sample

N = number of calibration curve points (typically, 3)

y = TCD area count of calibration species

x = volume fractions of calibration species

y_i = TCD area count of unknown species

β = slope of linear least squares fit to calibration points

The results of this analysis are included in all of the quantification graphs for each of the fuels examined in this program. This value was combined with the uncertainty of calibration gases and additional type B uncertainty and is presented as expanded uncertainty with a coverage factor of k = 2 in the data set and the appropriate figures in this report.

2.2.3.1 Gas Sample Storage System

The GC described above is a powerful and fast tool for analyzing gas samples. However, it only measures a single sample at a time. As described above data was taken primarily at two separate locations for this investigation and it is desirable to sample both locations simultaneously. To this end a gas sample storage system was designed to collect and store gas samples from the other discreet spatial point in the FSE for later analysis. The system in question is schematically illustrated in Figure 2.9.

The three primary valves are the Vici 6-Port switching valve, model number ED46UWE, the Vici 4-Port switching valve, model number ED44UWE, and the Vici 16 position switching valve, model number EMT4ST16MWE. These valves were sequentially switched in a particular order depending on the operation being conducted. The switching sequences was generically divided into two categories: Sample Acquisition and Sample Analysis.

A diagram of the sample storage system is presented in Figure 2.9, and the operation of the 4port and 6-port switching valves is illustrated in Figure 2.10. In sample acquisition the gas sample entered the sample port and flowed through the 4-port valve into a long coil of tubing, with a volume of ~100 mL. The sample then flowed through a 6-port valve and into the 16 position valve and 10 mL sample storage loop. After flowing through the sample storage loop the sample flowed through the 6-port valve again and out through the vent of the 6-port valve. Once a sufficient quantity of sample had flowed so that the sample storage loop was full of the sample, generally 2 minutes for this particular apparatus, the 6-port valve vent was closed and the 4-port valve switched to pressurize the sample still stored in the tubing coil and the sample storage loop. This step ensured that each sample was stored at the same pressure, placing the apparatus in an oven ensured each sample was stored at the same temperature. A summary of the control settings for capturing and analyzing gas samples from this system is provided in Table 2.4.

To retrieve the samples for acquisition the coil was first flushed with helium. The 16-position valve switched to the loop which was to be analyzed and the pressurized helium pushed the sample out of the loop and into the 6-port switching valve which fed the sample directly to the GC. At this point there was no flow and the sample was pressurized in the line upstream of the GC. The internal GC sample pump was used to draw the sample into the GC for analysis. The timing of the internal GC pump was critical because there was helium pressurized both upstream and downstream of the sample and the GC pump must act just long enough to ensure that only the sample gas was injected into the various columns of the GC. For the specific equipment used here there were two sample pumps with a capacity of 250 ml/min in the GC and it was found that running them for 15 s ensured that only the sample gas was injected into the columns. This time can vary based on the GC being used, the length and diameter of tubing and the pressure of the helium. After the sample was analyzed the 6-port valve switched to vent the remaining sample and purge the sample loop with helium. Then the 6-port valve switched back to the GC, the 16-position valve advanced to the next loop and the procedure was repeated until all of the samples had been analyzed. A summary of the control settings for capturing and analyzing gas samples from this system is provided in Table 2.4.

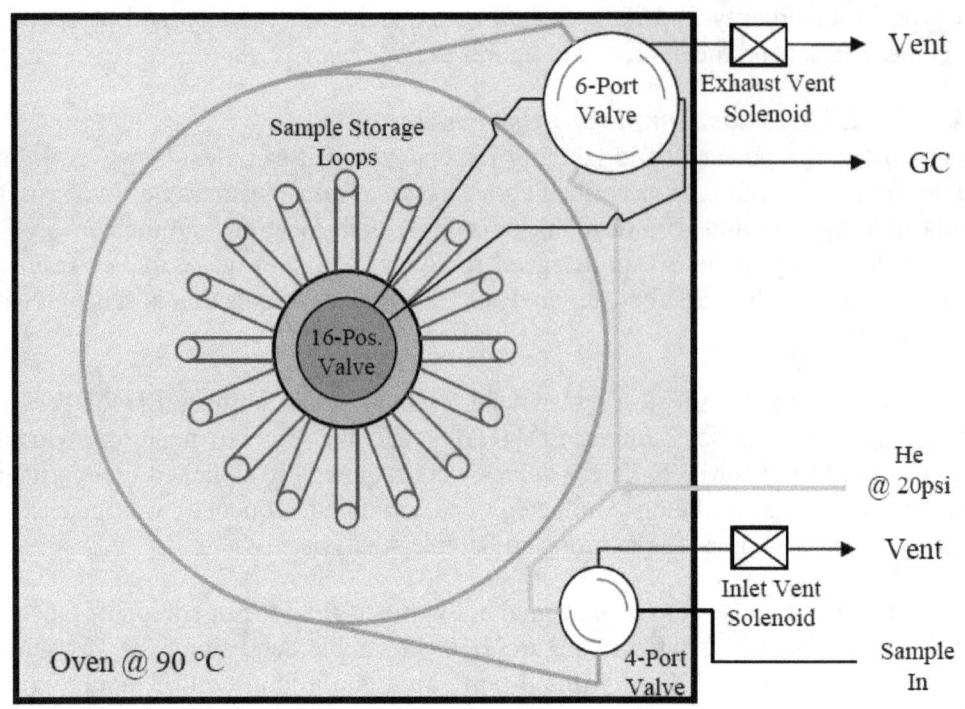

Figure 2.9: Schematic diagram of gas sample storage system in position B-A.

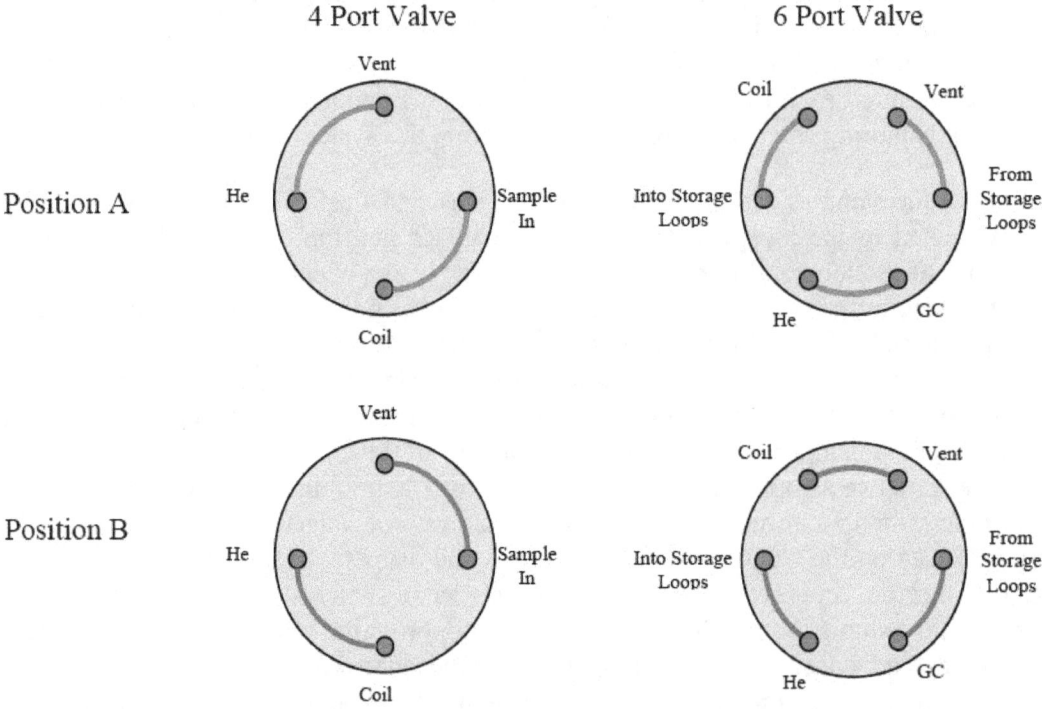

Figure 2.10: Positions of the control valves for the gas sample storage system.

Table 2.4: Sequence of controls for storing and then analyzing gas samples in the gas sample storage system.

Operation	Δt	4 Port	6 Port	16 Position	Exhaust Vent Solenoid	Inlet Vent Solenoid
Sample Collection (sample number N)						
Flow Sample	120s	A	A	N	closed	open
Pressurize Sample Loop	10s	B	A	N	open	closed
Close Loop, Flush Coil	10s	B	A	$N+1$	open	open
Analyze Sample (sample number N)						
Send Sample	240s	B	B	N	closed	closed
Flush Sample Loop	10s	B	A	N	closed	open

2.2.4 Soot Samples

2.2.4.1 Gravimetric

A gravimetric sampling system (shown in Figure 2.11) was used to measure soot mass fractions at the two sample locations within the enclosure. The design of the soot probe was similar to the gas sampling probes except the inner diameter of the sample tube was 6.4 mm. The soot sampling probes were conditioned with 65 °C water flowing at 1.0 L/min. A three way solenoid valve was used to rapidly switch from the bypass to sample flow. A sample gas mass flow rate of 2.75 standard L/min (N_2 @ 0 °C, 101.3 kPa) was drawn through the collection filter for a period of 60 s to 300 s. The collection filter was a 47 mm round Zeflour membrane filter with an aerosol retention efficiency of 99.99 % for 2 μm sized particles. A gas correction factor was applied to the mass flow rate measurement to account for the gas composition in the enclosure. The amount of time for sampling was determined by monitoring the pressure drop across the filter to ensure an optimal amount of filter loading.

The collection filters (shown at the base of the probes in Figure 2.11 below) and probe cleaning pads were conditioned in a desiccant drier before and after the tests. The conditioned filters were weighed using an analytic mass balance with an expanded uncertainty of 0.12 mg. After each soot sampling period, the probe was cleaned twice with gun cleaning pads. The total soot mass collected on both the filter and 2 cleaning pads was used in determining the soot mass fraction. Both the soot mass and sample mass flow rates were measured on a dry basis. For most of the tests conducted in this series between 10 mg and 200 mg of soot was collected during the 1 min to 5 min sample time. The extracted gas volume was corrected for the water removed by the method described in Sec. 0. The combined expanded relative uncertainty of the soot mass fraction measurement (for mass fraction measurements greater than 0.001 g/g) was in the range of 2 % to 5 % based on a propagation of uncertainty analysis.

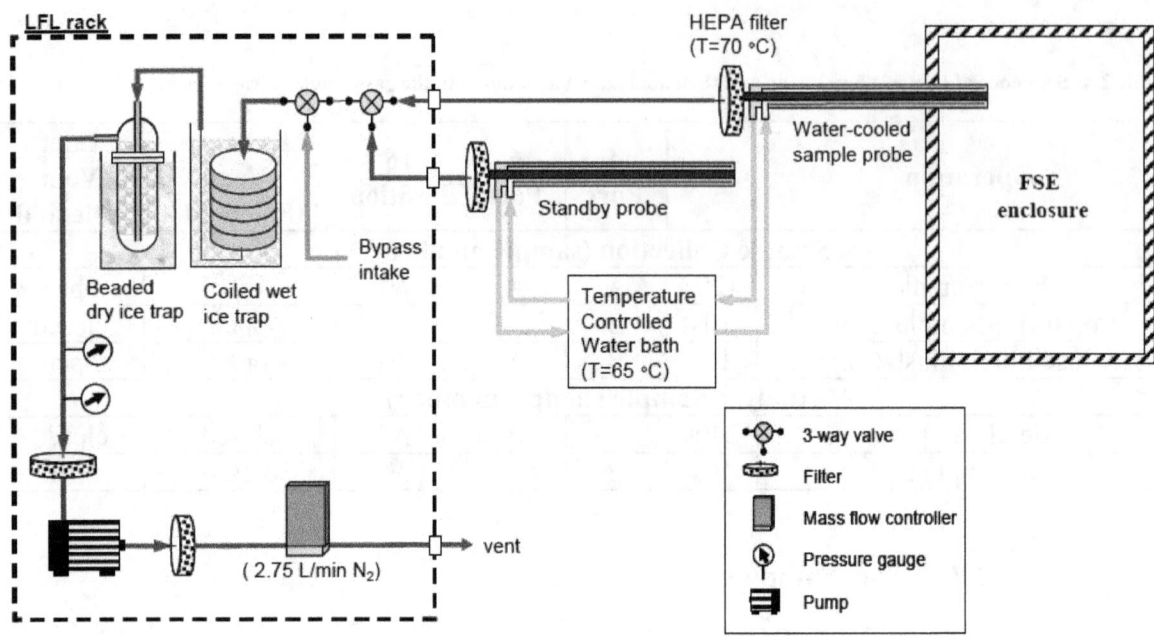

Figure 2.11: Schematic drawing of gravimetric soot sampling system.

2.2.4.2 Real time extractive

The real-time extractive soot probe was designed to measure the local soot concentration at a single location inside the enclosure using a light extinction method [30]. The probe continuously extracted a sample through an optical path; the measured attenuation of a laser passing through the sample was used to determine the soot mass concentration. Sample extraction was performed using a rack identical to the one shown schematically in Figure 2.11 by the dashed box labeled "LFL rack". A schematic of the probe is shown in Figure 2.12. The main section of the probe was constructed from two coaxial stainless steel tubes with a cap welded to each end, forming a seal around the annulus between the tubes. The sample was drawn through the inner tube, past the measurement section and out through a particulate filter. The inner tube had an outer diameter of 2.5 cm (1 in), a wall thickness of 3.2 mm (0.125 in) and was constructed from porous 316 stainless steel. The outer tube had an outer diameter of 5.1 cm (2 in), a wall thickness of 3.0 mm (0.120 in) and was constructed using 316 stainless steel. The sample gas flow rate was controlled by a mass flow controller (MFC) calibrated for N_2. The set point of the MFC was held fixed at 4 standard L/min (N_2 @ 0 °C, 101.3 kPa) for all tests. The actual flow rate was then obtained by correcting for the time-dependent local gas concentrations at the probe inlet based on the gas species analysis presented in Section 3.5.

Thermophoretic soot deposition was avoided by a nitrogen purge which flowed into the annular region between the tubes and was forced through the porous inner tube. The porous steel tube had a fine pore size (99.9 % collection efficiency for 0.2 µm particles) to maximize the pressure drop of the purge gas and distribute the flow as evenly as possible through the tube wall. The pressure drop associated with a typical 3.2 L/min N_2 purge flow was calculated to be approximately 1 kPa. The corresponding velocity at the inner tube surface was 0.7 mm/sec. The section of probe that was exposed to the heated environment inside the enclosure was protected by water cooling consisting of 6.4 mm (0.25 in) copper tubing wrapped around the exterior of the probe and a layer of the same ceramic fiber blanket used to cover the walls. House water (15 °C) flowed through the copper tubing at a rate of 0.95 L/min during the tests. The overall probe length was 1.22 m; the probe was inserted 61 cm (24 in) into the enclosure. The probe was positioned near the rear gas sampling probe (location specified in Table 2.5) so that sample conditions (temperature, species concentrations) could be determined.

The measurement section was located 1.12 m (44 in) from the sample inlet point. Two 1.9 cm (0.75 in.) O.D. tubes were welded to the probe to provide access for the laser and detector. Each of these tubes was connected to commercially available stackable lens tubes that were used to align the laser and detector and to shield the detector from stray light. A low-flowing nitrogen purge was used to prevent combustion products, including soot, from entering the tubes. The optical path was defined by the spacing of the optical tubes, which were flush with the inner diameter (1.9 cm) of the porous sample tube. The beam from a 780 nm diode laser was passed through a beam splitter to two identical high-speed silicon photodiodes. The detectors were sensitive to the range between 350 nm – 1100 nm. One detector was utilized to detect the attenuated signal that passed through the sample, the other compensated for the laser drift. Several optical components were placed in front of each detector. A narrow bandpass filter, centered at 780 ± 2 nm with FWHM of 10 ± 2 nm, acted to attenuate radiation other than that from the laser. A ground glass diffuser was used to expand the laser beam. Each detector was water cooled by coiled 3.2 mm (0.125 in) copper tubing (flowing 15 °C house water) wrapped in a layer of ceramic fiber blanket.

The determination of soot density (m_s) from an optical extinction measurement is based on Bouguer's law, in which the light extinction coefficient (K) is defined in terms of the attenuated intensity (I), and the reference intensity (I_o) of monochromatic light passing through a homogeneous smoke path of distance (L) [31]:

$$K = -\frac{\ln(I/I_0)}{L} \qquad 1$$

$$m_s = \frac{K}{\sigma_s} \qquad 2$$

where σ_s is the mass specific extinction coefficient. The recommended value of σ_s for flame generated smoke in over-ventilated fires is 8.7 m²/g ± 5.4 % (standard relative uncertainty) [31]. This value is based on measurement at 632 nm. It has been shown that the mass specific extinction coefficient is approximately inversely proportional to wavelength [32]. For this reason, the present analysis uses a value of σ_s = 6.96 m²/g.

The expanded combined uncertainty of the soot mass concentration was computed as the sum of the individual standard uncertainties [u(x$_i$)] associated with each of the terms that influence the soot mass measurement. Drift in the laser baseline, as measured by pre- and post-test values of I/I$_o$, was the dominant contributor to the overall uncertainty. In some cases this term was small, on the order of 1 – 3%, however it ranged as high as 19%. The average was approximately 8%. The combined uncertainty of the optical soot density measurement was calculated as follows:

$$U_{m_s} = 2\sqrt{\left(\frac{\partial m_s}{\partial \ln(I/I_0)}\right)^2 u(\ln(I/I_0))^2 + \left(\frac{\partial m_s}{\partial L}\right)^2 u(L)^2 + \left(\frac{\partial m_s}{\partial \sigma_s}\right)^2 u(\sigma_s)^2} \qquad 3$$

The estimate of the combined relative expanded uncertainty was 30%.

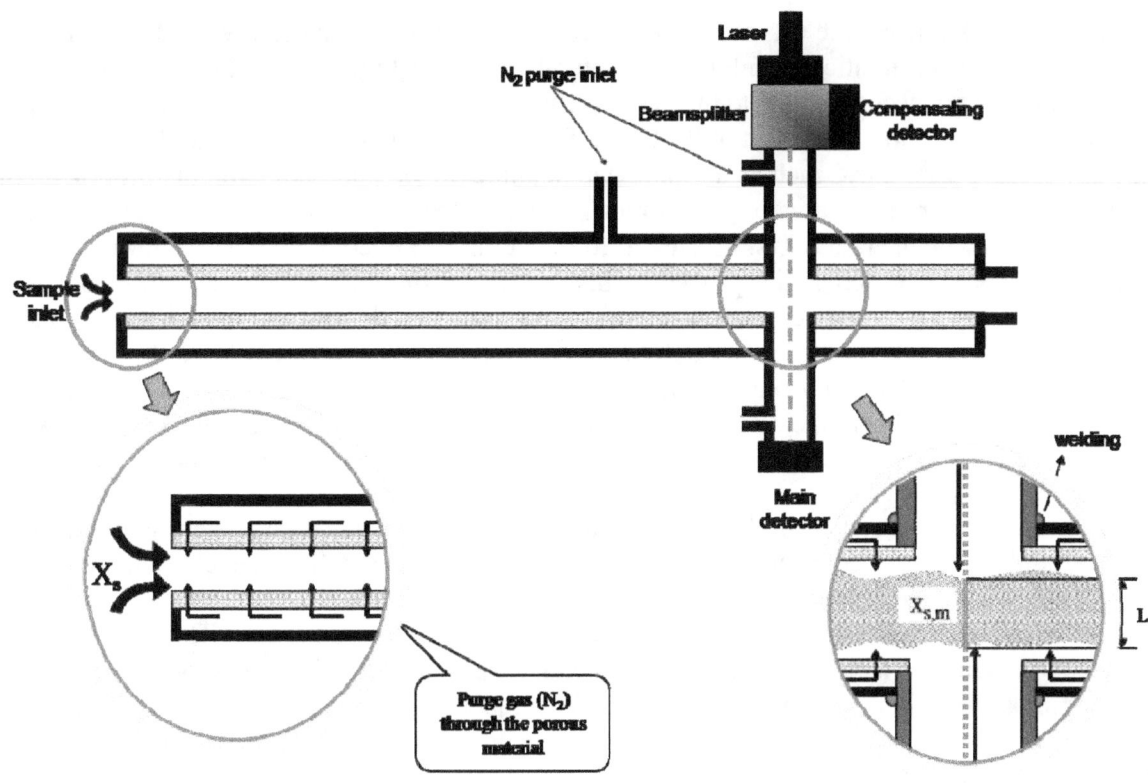

Figure 2.12: Schematic of real time extractive soot measurement probe.

2.2.5 Thermocouples

2.2.5.1 Aspirated Thermocouples

A bare-bead thermocouple situated in a compartment fire typically experiences radiative exchange with walls, hot smoke, flames, and the surrounding environment with the effect that the measured temperature is not the true gas temperature. Accurate correction for these effects is complex, due to temporally and spatially varying local temperatures, velocities, and species. To reduce the effect of energy exchange on temperature measurement accuracy, aspirated thermocouple probes were used in addition to bare-bead thermocouples in this study.

An aspirated thermocouple probe is a bare-bead thermocouple contained within a small cylindrical metal tube through which the sample gas flows. If the flow over the bead is at least 5 m/s, a more accurate gas temperature measurement may be obtained [33]. According to Blevins [34], higher flows may be required depending on the thermal environment. Aspirated thermocouple probes may be shielded by a single cylindrical tube or by two or more concentric cylindrical tubes. In either case, the flow and thermal conditions and the detailed design of the assembly can impact measurement accuracy. Double-shielded aspirated thermocouple probes based on a design from NACA, the predecessor agency to NASA, were used in this study [35]. Figure 2.13 shows a drawing of an end-hole type NACA design aspirated thermocouple probe. Models with the entrance hole perpendicular to the probe axis were also used.

Each aspirated thermocouple was connected directly to a venturi pump using 9.5 mm (3/8 in) OD copper tubing. A schematic of venturi pump setup is shown in Figure 2.14. Filtering and drying of the flow was not necessary due to the design of the venturi pump which merely blows the flow out into the hood. Flows were set at 24 L/min for each aspirated probe. While the volumetric flows were set with flow meters at room temperature to be the same for all probes, high temperature compartment gases produced much higher velocities at the bead compared to those produced by low temperature gases. The uniform setting of the cold volumetric flows kept the mass flows consistent across the probes. This velocity difference effect was not completely proportionate to the gas temperature differences since a higher flow would experience a greater pressure drop and flow resistance through the probe and tubing. Due to the large flows pumped through the aspirated thermocouple probes, the resulting temperature represents a volumetric average over a several centimeter diameter region at the end of the probe. For further discussion of the probe and gas interaction have been previously reported [5].

Each aspirated thermocouple probe was attached to the data acquisition system using K-type thermocouple wire and connectors. During each experiment, the flow meters and measured temperatures were monitored. These checks were performed in order to determine if any probe system became clogged so it could be unclogged with high pressure air. The difference in temperature signal between an inoperative probe and a properly flowing probe was obvious. A functioning aspirated thermocouple showed higher frequency temperature fluctuations due to the transient thermal environment and effective convective heat transfer while a non-functioning probe would not show rapidly fluctuating temperatures since the large mass of hot metal of the probe radiating to the bead and lack of convection would dampen any short fluctuations. A probe typically required about 1 min when activated to overcome accumulated heat and reach the true gas temperature.

To evaluate measurement uncertainty and instrument time response, a series of detailed flow and heat transfer calculations, focusing on double-shielded aspirated thermocouples and bare-bead thermocouples were performed for a previous study. A detailed description of the calculations and results can be found in Ref. [5]. It was determined that in a worst case scenario, with a gas temperature of 1200 K and an ambient temperature of 300 K, the uncertainty of the aspirated thermocouple measurement was 25 %. However, since the aspirated thermocouples did not 'see' a large temperature difference, the uncertainty of the aspirated thermocouple temperatures were much less than that, typically less than 10 %. The aspirated thermocouples do however have a larger temporal uncertainty and tend to average temperatures over time and may miss fast gas temperature fluctuations due to the thermal mass of the radiation shields.

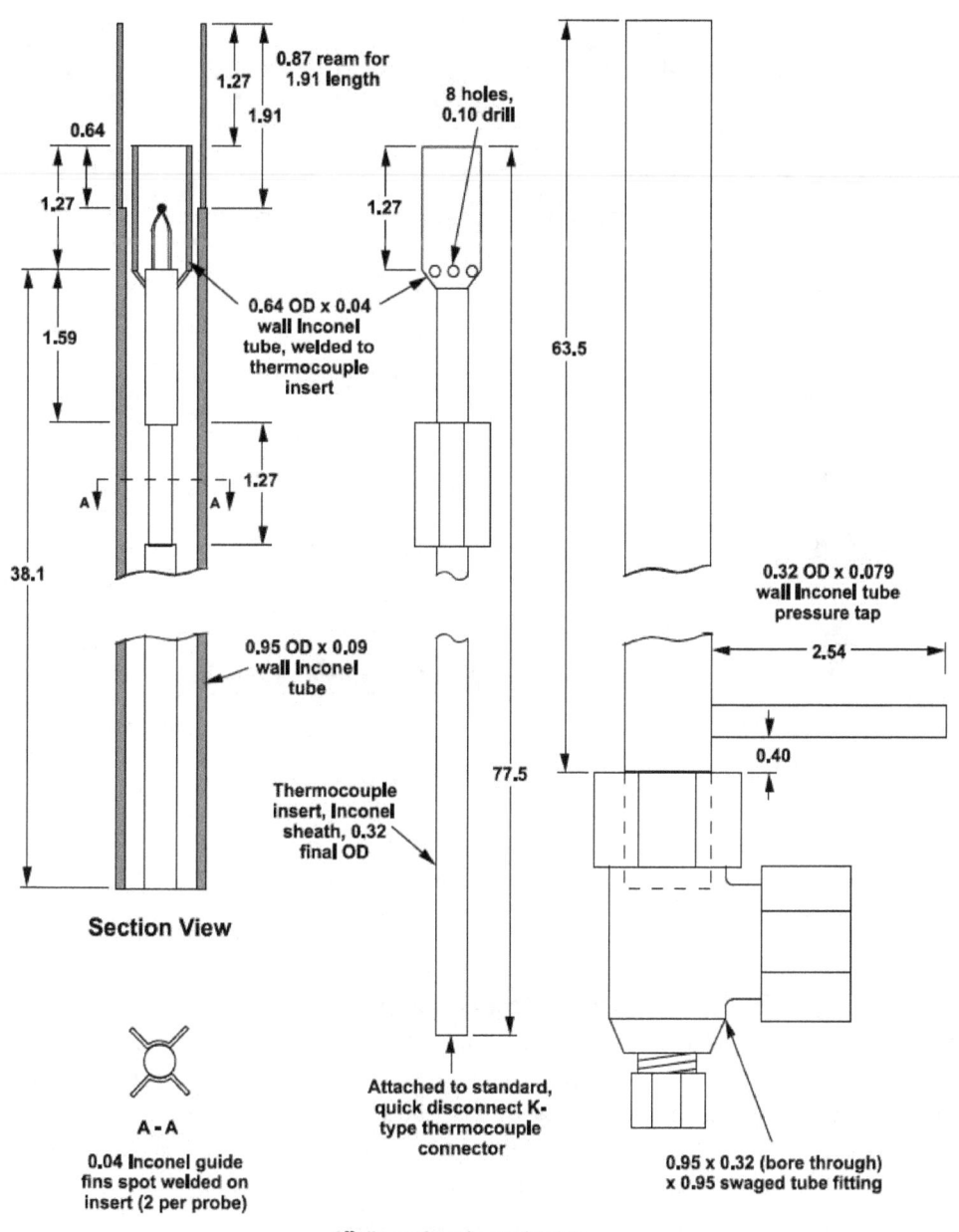

Figure 2.13: Detailed drawing of aspirated thermocouple using NACA design [34].

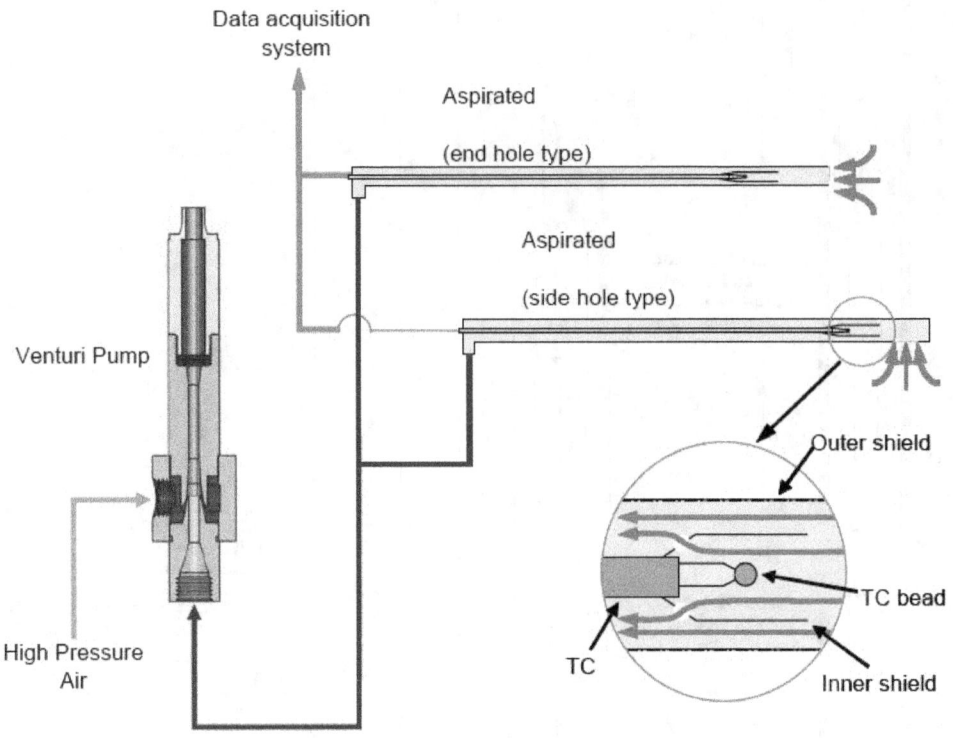

Figure 2.14: Schematic drawing of aspirated thermocouple measurement hardware.

2.2.5.2 Radiation Effects on Bare Beads

Because bare bead thermocouples were used for some measurements inside the room it is important to discuss the effect of radiative losses on the value of this measurement. Unlike the aspirated thermocouples which are specifically designed to eliminate radiative losses from the measurement [34] the bare bead thermocouples are subject to radiative losses. This occurs for example when you have an optically thin flame, e.g. premixed methane, in which the temperature is being measured by a thermocouple and the surrounding ambient conditions are much colder than the flame, e.g. 2000 K flame in a 300 K room. In some of our tests, the thermocouple environment is optically thick due to heavy soot loading, and the thermocouple does not radiatively 'see' a cool surface, such as the vent of the FSE. However there are some cases where the thermocouples may be reading very high temperatures in a optically thin environment with significant radiative exchange through the door with the ambient room conditions.

Consider the example where there is a thermocouple located on the center of the front thermocouple tree. That thermocouple would be located at a position of 120 cm x 72 cm x 120 cm in room coordinates. That places it 72 cm from the center of a door that is nominally 40 cm wide by 200 cm tall. Now consider that the thermocouple reads 1200 °C, the temperature outside the room is 20 °C and the environment within the room is optically thin. The temperature being read by the thermocouple is not the actual temperature, but rather a steady state balance between heat entering the thermocouple by convection and heat leaving the thermocouple by radiation. The general expressions of heat transfer in this case, for convection and radiation, respectively, are:

$$q_{conv} = A_{conv} h (T_{gas} - T_{bead}) \qquad 4$$

$$q_{rad} = A_{rad} \sigma \varepsilon (T_{bead}^4 - T_{cold}^4) \qquad 5$$

The temperature that is desired is the actual temperature of the gas, T_{gas}, the thermocouple bead temperature, T_{bead}, and the cold surrounding temperature, T_{cold}, are presumably known. One can then solve for the actual temperature of the gas:

$$T_{gas} = T_{bead} + \frac{A_{rad} \sigma \varepsilon (T_{bead}^4 - T_{cold}^4)}{A_{conv} h} \qquad 6$$

Since updating the temperature of the gas can change the convective heat transfer coefficient, h, and the emissivity, ε, so that the equation should be solved iteratively until a correct gas temperature is obtained. One additional complication here is that the area subject to radiation is not the same as the area subject to convection, $A_{conv} \neq A_{rad}$. If the rest of the room should be approximately the same temperature as the gas at the bead then the thermocouple in this situation only sees a small angle of the cold environment due to the size and location of the doorway. An analysis of this problem with the given geometry showed that the temperature change for this worst-case scenario of a thermocouple bead inside an enclosure was only 2 % at 1200 °C. By contrast, in a worst case scenario, if the bare bead thermocouple were at 30 °C, seeing an environment at 1200 °C, the uncertainty in the bare bead temperature reading could be on the order of 200 % [5]. Taking into account the other components of uncertainty including random variations and the inherent accuracy of the thermocouple a combined expanded uncertainty of -20 % to +6 % with a coverage factor of 2 is reported in Table 2.6.

2.2.6 Heat flux gauges

Total heat flux was measured at six locations during each experiment. The heat flux gauges were 6.4 mm diameter Schmidt-Boelter type, water cooled gauges with embedded type-K thermocouples. The particular model information is contained in Appendix D. The nominal range for the gauges was 150 kW/m². Schmidt-Boelter gauges measure a temperature difference across a thin insulating material using a thermopile to generate a voltage from the small temperature difference. These gauges typically generate voltages much less than 100 mV even for heat fluxes near their maximum range.

Each gauge was inserted in the floor flush with the upper surface and facing vertically upward. The floor heat flux gauges were located in three places, just outside the doorway on the centerline of the floor and straddling the burner at y = 90 cm and y = 270 cm. The exact location coordinates for the gauges are listed in Table 2.5. The condition of the installed gauges was checked periodically. If significant soot accumulated on a gauge, it was brushed off. If a gauge was no longer flush with the surface of the floor, a note was made, but there was no attempt to move the gauge since the gauges were very difficult to access.

Heat fluxes as high as 300 kW/m^2 were observed. These heat fluxes are beyond the stated range of the gauges. According to the manufacturer, the calibrations remain linear and valid beyond the stated range as long as the materials do not degrade and change the sensitivity of the gauge. Previously the calibration of the gauges has been checked after experiencing these large heat fluxes [5]. Each gauge's responsivity was found to remain within 3 % of the factory calibration.

The main sources of uncertainty related to the total heat flux measurements are: the calibration, soot and dust deposition, and shifting of the gauge surface below the floor. These sources will be described and the total uncertainty estimated for the reported measurements. A model of uncertainty for heat flux gauge measurements in fire environments can be found in the study by Bryant et al. [36].

The total heat flux gauge calibration from the manufacturer was used to convert millivolt readings to kW/m^2. This calibration was performed using cooling water at 23 °C ± 3 °C. The cooling water in the Large Fire Laboratory was found to be within the same range. The manufacturer reported a ±3 % expanded uncertainty in the responsivity (the slope in kW/m^2/mV). Calibrations at the NIST facility have varied within the 3 % range of the nominal manufacturer's calibration. A recent round-robin study of heat flux gauge calibration consistency [38] sent the same heat flux gauges to multiple laboratories around the world and found that while several calibrations fell within the 3 % range, if some outlier data were included, then the uncertainty rose to around 8 %. For this current project, an uncertainty of ±6 % for gauge calibration was chosen as fairly conservative since the NIST calibration was within the 3 % range in the round-robin study.

While the cooling water was supplied at approximately 23 °C, the fire heated the water such that the gauge temperature typically rose to between 40 °C and 60 °C, and less frequently to 100 °C. For the fires where the water temperatures increased to between 40 °C and 100 °C, the heat fluxes were on the order of 100 kW/m^2 to 300 kW/m^2 which represent blackbody temperatures in the 950 °C to 1300 °C range. The most extreme combination (affecting uncertainty) of cooling water and environment temperature would be a 75 °C increase in cooling water in a 950 °C environment. This combination would only have about a 0.5 % effect on the measured heat flux. The effect was determined by calculating the ratio of the T^4 difference between 950 °C and the 25 °C cooling water with 950 °C and the 100 °C cooling water. This is a simplified comparison which assumes everything else is equal, but generates an approximation of the magnitude of the cooling water effect under specified conditions.

Heat flux uncertainty due to soot and dust deposition is difficult to quantify. For many tests, such as those burning methanol, ethanol, and natural gas, there was little to no contact with soot or combustion products. Also, even for the sootier fuels at low HRRs, the lower layer remained as air with little opportunity for soot-laden gases to contact the gauges. For those experiments with sooty fuels and under-ventilated conditions, combustion products including soot sometimes impinged on the gauges. For these periods of time, it was estimated that the soot coating on the gauge would add an additional uncertainty of ±10 % due to variations in surface emissivity, and soot agglomerates shadowing the surface of the gauge.

The physical shifting of the gauge surface below the floor could have impact on a heat flux measurement if the solid angle viewable by the gauge was significantly diminished. Since the gauge is not sensitive either in calibration or application to radiation at angles close to the plane of the gauge surface due to reflection, and the radiation approaching from the lowest angles is generally from the coolest regions of the enclosure, the gauge would have to be below the surface of the floor by a few millimeters or more for there to be a significant impact on its measurement. Neither gauge was ever observed to be shifted by that amount in the course of testing.

2.3 Sampling locations

The gas species, temperatures, heat fluxes, and soot data were taken at various discrete places in the room. Table 2.5 lists the various data that was acquired in the room and where each probe was located. In the case of the movable probe, the vertical location (z – direction) was variable between z=100 cm and z=240 cm. In the data files published online as a part of this report the position of the movable probe is noted, in centimeters, for each of the cases where the movable probe was used. In some of the cases the relative location of a particular probe is indicated by it's data label. As an example, TFSampA refers to the temperature measurement at the front of the room; TF30 refers to the temperature measurement at the front thermocouple tree, located 30 cm from the floor.

Table 2.5. Location of measurement probes inside of the enclosure.

Probe Description	Data Label	x (cm)	y (cm)	z (cm)
O₂ at Rear Sampling location (Rack #1)	O2Rear	189	286	208
CO₂ at Rear Sampling location (CO signal) overrange	CO2Rear	189	286	208
CO at Rear Sampling location	CORear	189	286	208
UH at Rear Sampling location (before GC)	UHRear	189	286	208
Rear Temperature at Sampling Location PtRh	TRSampPtRh	169	286	208
O₂ at Front Sampling location (Rack #2)	O2Front	189	25	208
CO₂ at Front Sampling location	CO2Front	189	25	208
CO at Front Sampling location	COFront	189	25	208
UH at Front Sampling location (before auto sample storage)	UHFront	189	25	208
Front Temperature at Sampling Location PtRh	TFSampPtRh	169	25	208
O₂ at Moving Sampling location (Rack #3)	O2Move	120	290	z
CO₂ at Moving Sampling location	CO2Move	120	290	z
CO at Moving Sampling location	COMove	120	290	z
Moving Sample in room Aspirated TC	TRMoveSamp	120	290	z
Total Heat Flux Gauge Rear Floor A (SN=131835)	HFRFL	119 5	266	0
Total Heat Flux Gauge Front Floor B (SN=131836)	HFFFL	119 5	90	0
Total Heat Flux Gauge Outside Floor C (SN=131833)	HFOFL	119 5	-20	0
Total Heat Flux Gauge Rear Ceiling D (SN=131837)	HFRCE	119 5	266	233
Total Heat Flux Gauge Center Ceiling E (SN=131838)	HFCCE	119 5	178	233
Total Heat Flux Gauge Front Ceiling F (SN=131834)	HFFCE	119 5	90	233
Temperature of Total Heat Flux Gauge Rear Floor A	THFRFL	119 5	266	0
Temperature of Total Heat Flux Gauge Front Floor B	THFFFL	119 5	90	0
Temperature of Total Heat Flux Gauge Outside Floor C	THFOFL	119 5	-20	0
Temperature of Total Heat Flux Gauge Rear Ceiling D	THFRCE	119 5	266	233
Temperature of Total Heat Flux Gauge Center Ceiling E	THFCCE	119 5	178	233
Temperature of Total Heat Flux Gauge Front Ceiling F	THFFCE	119 5	90	233
Interior Enclosure Surface Temp Near Total HF Gauge Rear Floor A	TSHFRFL	119 5	266	0
Interior Enclosure Surface Temp Near Total HF Gauge Front Floor B	TSHFFFL	119 5	90	0
Interior Enclosure Surface Temp Near Total HF Gauge Outside Floor C	TSHFOFL	119 5	-20	0
Interior Enclosure Surface Temp Near Total HF Gauge Rear Ceiling D	TSHFRCE	119 5	266	233
Interior Enclosure Surface Temp Near Total HF Gauge Center Ceiling E	TSHFCCE	119 5	178	233
Interior Enclosure Surface Temp Near Total HF Gauge Front Ceiling F	TSHFFCE	119 5	90	233
Exterior Enclosure Surface Temp Near Total HF Gauge Rear Floor A	TSXHFRFL	119 5	266	0
Exterior Enclosure Surface Temp Near Total HF Gauge Front Floor B	TSXHFFFL	119 5	90	0
Exterior Enclosure Surface Temp Near Total HF Gauge Rear Ceiling C	TSXHFRCE	119 5	-20	0
Exterior Enclosure Surface Temp Near Total HF Gauge Center Ceiling D	TSXHFCCE	119 5	266	233
Exterior Enclosure Surface Temp Near Total HF Gauge Front Ceiling E	TSXHFFCE	119 5	178	233
TC Tree Rear TC 1 in up	TR3	120	288	2 5
TC Tree Rear TC 1 ft up	TR30	120	288	30
TC Tree Rear TC 2 ft up	TR60	120	288	60
TC Tree Rear TC 3 ft up	TR90	120	288	90
TC Tree Rear TC 3 5 ft up	TR105	120	288	105
TC Tree Rear TC 4 ft up	TR120	120	288	120
TC Tree Rear TC 4 5 ft up	TR135	120	288	135
TC Tree Rear TC 5 ft up	TR150	120	288	150
TC Tree Rear TC 6 ft up	TR180	120	288	180
TC Tree Rear TC 7 ft up	TR210	120	288	210
TC Tree Rear TC 7 ft 11 in up	TR237	120	288	237 5
TC Tree Front TC 1 in up	TF3	120	72	2 5
TC Tree Front TC 1 ft up	TF30	120	72	30
TC Tree Front TC 2 ft up	TF60	120	72	60
TC Tree Front TC 3 ft up	TF90	120	72	90
TC Tree Front TC 3 5 ft up	TF105	120	72	105
TC Tree Front TC 4 ft up	TF120	120	72	120
TC Tree Front TC 4 5 ft up	TF135	120	72	135
TC Tree Front TC 5 ft up	TF150	120	72	150
TC Tree Front TC 6 ft up	TF180	120	72	180
TC Tree Front TC 7 ft up	TF210	120	72	210
TC Tree Front TC 7 ft 11 in up	TF237	120	72	237 5
Interior Surface Temperature Back Wall Centerline Top	TSBWCTop	125	356	180
Interior Surface Temperature Back Wall Centerline Middle	TSBWCMid	125	356	120
Interior Surface Temperature Back Wall Centerline Bottom	TSBWCBot	125	356	60
Exterior Surface Temperature Back Wall Centerline Top	TSXBWCTop	125	360	180
Exterior Surface Temperature Back Wall Centerline Middle	TSXBWCMid	125	360	120
Exterior Surface Temperature Back Wall Centerline Bottom	TSXBWCBot	125	360	60

2.4 Data acquisition

Data acquisition (DAQ) for this series of experiments was divided into two systems. One DAQ system was dedicated to fuel flows, oxygen depletion calorimetry, and the constituent measurements required to calculate heat release rate using that method. The other DAQ system was used to record signals from all other measurements (refer to previous report). Each DAQ system used National Instruments hardware and was controlled with LabVIEW software. The calorimetry DAQ system has been previously described in detail [29].

For this series of experiments, the channel list contained in Appendix C was used to program the DAQ system. The types of measurements included: gas analyzers, dew point readers, heat flux gauges, pressure transducers, and thermocouples. These measurements were recorded on the DAQ hardware as voltages with 200 samples recorded every second. Each second, the average value for each channel was then converted to meaningful physical units. Two event marking channels were used to note the time of important events such as ignition, fuel flow change, or extinguishment. These event marker channels, which are in both DAQ programs, were especially useful in synchronization of the two data sets.

The data acquisition hardware had 16 bit precision, with stated accuracies of the data acquisition board and multiplexing module equal to 0.014 % and 0.015 % of the reading. These uncertainties were orders of magnitude lower than those from other sources in all of the measurements reported here.

2.5 Data post-processing

A Matlab script file was created for post-processing all data files generated during the test series. This program was used to make corrections to the data, generate plots, and save results to ASCII text files for archival purposes. The program was also used to compute time averaged values and uncertainties for examining trends in the data. An input file was used to allow batch processing of the raw data files. The input file contained the parameters needed for the heat release calculation (this file was also read by the DAQ program during the data collection process). Additional parameters were added to the end of the standard HRR input file to account for the gravimetric soot measurements and to record the time windows when channels had known missing or corrupted data.

The first step in data reduction was to inspect the data files and lab notebooks for erroneous data resulting from open channels, loss of sample flow, or some other instrument or data acquisition malfunction. Because data were collected on two separate computers, the series were synchronized to a common reference time. The ignition time was marked using a virtual event channel on each computer and defined as time zero for the reduced data. The gas analyzer measurements from inside the FSE and exhaust hood measurements were shifted in time to account for the sample flow transfer (delay) time as discussed in section 2.2.2.

Corrections to the heat release rate measurements were applied to account for the exhaust flow calibration factor and drift in the oxygen analyzer.

Since the gases sampled from the FSE were dried before entering the detectors, an estimate of the water removed can be made in order to correct the measurements to the *in situ* wet volume fraction. In this report the wet volume fraction of gases is only used to determine the mixture fraction values, see section 4.1. Other gas species measurements are reported on a dry basis. This is done because converting the dry gas sample volume fractions to a wet basis introduces an additional source of uncertainty. The general combustion reaction assuming all the fuel is reacted and that the soot can be represented as pure carbon is:

$$C_x H_y O_z + aO_2 \rightarrow bCO_2 + cCO + dCH_4 + eC + fH_2O \qquad 7$$

The molecular yield of water can be related to the combustion product yields using the known hydrogen/carbon (y/x) ratio of the fuel:

$$f = \frac{y}{2x}(b+c+d+e) - 2d \qquad 8$$

If the yield of soot is small compared to the other products, the water volume fraction, X_{H2O}, can be estimated from Eq.11.

$$X_{H_2O} = \frac{y}{2x}\left(X_{CO_2,wet} + X_{CO,wet}\right) \qquad 9$$

The relationships for wet CO_2 and CO are given by the following:

$$X_{CO,wet} = \frac{X_{CO,dry}}{1 + \frac{y}{2x}\left(X_{CO_2,dry} + X_{CO,dry}\right)} \qquad 10$$

$$X_{CO_2,wet} = \frac{X_{CO_2,dry}}{1 + \frac{y}{2x}\left(X_{CO_2,dry} + X_{CO,dry}\right)} \qquad 11$$

Other gas volume fraction measurements performed on a dry basis were corrected using the following relationship:

$$X_{spec,wet} = X_{spec,dry}\left(1 - X_{H_2O}\right) \qquad 12$$

The total hydrocarbons can contribute to the formation of water, however the gas composition measurements confirmed that when total hydrocarbons were present in significant quantities, they were in the form of unburned fuel (methane in the natural gas tests). Unburned fuel does not contribute to the formation of water. Therefore, the resulting relative error in the water volume fraction estimation due to neglecting hydrocarbons was always less than 3 %. The error in the water volume fraction estimate due to neglecting soot was as much as 10 % for the highly sooting fuels. However, since the soot measurements were sparse, we chose to report the results on a consistent basis. In some cases the volume fraction of hydrogen measured by GC was as high as 0.10 which may result in an additional uncertainty of the water volume fraction by as much as 10 %. A more accurate estimate of water volume fraction can be made for the short time windows where soot and hydrogen volume fractions were both collected.

2.6 Uncertainty

There are different components of uncertainty in the temperatures, total heat flux, soot mass fraction, chemical species, and heat release rate reported here. Uncertainties are grouped into two categories according to the method used to estimate them. Type A uncertainties are evaluated by statistical methods and type B are evaluated by other means [37]. Type B analysis of systematic uncertainties involves estimating the upper (+a) and lower (-a) limits for the quantity in question such that the probability that the value would be in the interval (±a) is 95 percent. After estimating uncertainties by either Type A or B analysis, the uncertainties are combined in quadrature to yield the combined standard uncertainty. Multiplying the combined standard uncertainty by a coverage factor of two results in the total expanded uncertainty that corresponds to a 95 percent confidence interval (2σ).

Components of uncertainty are tabulated in Table 2.6. Some of these components, such as the zero calibration elements, are derived from instrument specifications. Other components, such as radiation loss and thermophoretic soot deposition include past experience with these measurements. The uncertainty in the air temperature measurements does include radiative cooling which is likely to result in a measured temperature lower than the actual gas temperature. Smoke measurements were primarily conducted gravimetrically. Part of the uncertainty was attributed to the accuracy of the mass scale used to weigh the soot filters, see section 2.2.4.1, and part of the uncertainty was due to the flow rate which is measured after the gases have been cooled and is therefore highly depended on the temperature at the entrance to the soot probe. Additional uncertainty is introduced by the water correction discussed in section 2.5. Uncertainties in the heat release rate measurement can be traced to variations in the hood duct flow profile, soot and total hydrocarbons which are not accounted for, and a small instrument uncertainty. The general function of the heat release rate measurement is discussed in section 2.2.1 and the uncertainty of the measurement in this hood has been well documented [29, 38]. The associated uncertainty in the ideal heat release rate, used to calculate combustion efficiency is related primarily to the purity of the fuel and the uncertainty in the fuel flowrate measurement device, both of which are small. The gas analyzers, $CO/CO_2/O_2/THC$, use precision mixed zero and span gases and have a small uncertainty reported by the manufacturer. For these devices the random and mixing/averaging due to long sample lines are a larger source of uncertainty. The gas chromatograph was calibrated with a variety of different gas standards and some mixtures that were made in-house. The calibration of the GC for all of the species that it detects is a long procedure and was only conducted at the beginning and end of the test series which introduces significant uncertainty beyond the gas standard mixture uncertainty. Additionally the ANOVA analysis presented in section 2.2.3 provides information on the statistical uncertainty associated with using the GC. Because of the large amount of sample tubing and the sparse timing of GC data there is also an uncertainty in the GC sample timing. Generally the GC sample time can be considered to be ±30 s. The heat flux gauges used here were generally very precise devices and despite being used beyond their calibrated range have been shown to be quite linear and repeatable, cf. section 2.2.6. The larger sources of uncertainty came as a result of the cooling water being unable to remove heat from the heat flux gauge fast enough and because of thermophoretic soot deposition on the heat flux gauge window surface. The uncertainties are reported as a range in Table 2.6 and are represented by error bars in the associated figures.

Table 2.6: Uncertainty of measurements

	Component Standard Uncertainty	Combined Standard Uncertainty	Total Expanded Uncertainty

Temperature
    Calibration
    Radiative Cooling
    Random
	
	 ±1 %
	 -10 % to 0 %
	 ±3 %
		
		 -10 % to +3%
			
			 -20 % to +6 %

Heat Flux
    Soot Deposition
    Cooling Water Temp
    Calibration
    Random
	
	 -5 % to 0 %
	 ±5 %
	 ±1 %
	 ±3 %
		
		 -8 % to +6 %
			
			 -16 % to +12 %

Gas Analyzers
    Zero and Span Gas
    Equiptment Uncertainty
    Mixing and Averaging
    Random
	
	 ±1 %
	 ±1 %
	 ±5 %
	 ±3 %
		
		 ±6 %
			
			 ±12 %

Gas Chromatography
    Anova
    Calibration
    Mixing and Averaging
    Random
	
	 ±1 %
	 ±5 %
	 ±5 %
	 ±3 %
		
		 ±8 %
			
			 ±16 %

Soot Mass Fraction
    Mass Measurement
    Volume Flow Rate
    Water Estimation
    Random
	
	 ±1 %
	 ±1 %
	 ±2 %
	 ±3 %
		
		 ±3 %
			
			 ±5 %

Measured Heat Release Rate
    Exhaust Flow Rate
    Soot and THC
    Instruments Uncertainty
    Random
	
	 ±5 %
	 ±3 %
	 ±1 %
	 ±3 %
		
		 ±7 %
			
			 ±14 %

Ideal Heat Release Rate
    Fuel Purity
    Equiptment Uncertainty
    Random
	
	 ±1 %
	 ±1 %
	 ±3 %
		
		 ±3 %
			
			 ±6 %

3 RESULTS

3.1 List of Test Conditions

Thirty experiments were conducted in the full scale enclosure and are listed in Table 3.1. The first column is the Test ID used to identify each test in this report; the second column is the date at which the test was completed followed by the fuel consumed. Next the quantity of fuel consumed, when a fixed quantity of fuel was used, is listed followed by the number of burners, the size of each burner and the doorway size, in terms of fractions of a standard 80 cm doorway. Whether the room flashed over or not (determined by oxygen depletion in the upper layer), the duration of the test and the actual ignition time are also listed. A general discussion of the test conditions for each test follows.

In tests ISONG1, ISONG2, ISONG3, ISOHept4, and ISOHept5 a slightly different room configuration to that described above in section 2.1 was used. The same ceramic fiber blanket insulation was used with similar ceramic anchors except that only one 2.5 cm layer of insulation was used instead of 5 cm. Also, the exterior skin of the room was constructed of gypsum board instead of steel. The internal dimensions of the room were slightly smaller as well. The internal dimensions for these tests were 239 cm x 239 cm x 360 cm with the same uncertainty of ± 2 cm in each direction. From Table 3.1 it can be seen that these tests were conducted almost 5 months earlier than the subsequent tests. These were the first under-ventilated FSE tests in this test series and therefore they were conducted primarily to evaluate the robustness of the structure, the instrumentation, and the burner. For these tests a special burner was also tested, the water cooled, pump fed, variable surface area burner described in section 2.1.4 was used. It was hoped that this burner could be used to control the heat release rate in the room while maintaining a pool fire type scenario by utilizing the variable surface area feature of the burner. For the natural gas cases the burner was filled with gravel and the flow of natural gas was metered to the burner.

This burner worked well for natural gas, however no better than other simpler gravel burners which are already used in the lab. Figure 3.1 shows the measured and ideal heat release rate for test ISONG3. Here we can see that the measured heat release rate increases at nearly the same time as the change in flow and both the measured and metered heat release rates remain fairly steady at the same time. For heptanes, the gravel was removed and the flow rate of heptanes was also metered along with a measurement of the pool height and therefore surface area by means of the pressure transducer discussed previously. The success of controlling the heat release rate for the heptanes fuels with this burner was significantly more limited. Figure 3.2 presents the measured and metered heat release rates for test ISOHept5. The metered fuel flow rate was controlled very carefully and several steps of steady pump flow rate were tested. However, as Figure 3.2 illustrates the measured heat release rate from the fire was not steady at any point during the test despite the steady flow rates. This instability was assumed to be a result of a competition between the burning rate (a function surface area, and net heat flux) and manually controlled fuel delivery rate. The process could not be well controlled and it was decided that this burner was not adequate for the tests that needed to be done. Additionally, the gypsum walls of the structure failed and the structure was unusable after test ISOHept5. Thus testing was halted and the structure discussed above in section 2.1 was constructed and used for the subsequent experiments. Tests 6 and 7 do not exist and were skipped because of a numbering error. The numbering was not corrected here in order to maintain the persistence of the test names that are recorded permanently in various places including laboratory notebooks, experimental logs, and computer data files.

Table 3.1: List of test conditions considered in this report.

Test ID	Date	Fuel	Fuel Mass (kg)	# of Burners	Burner Size (m^2)	Doorway (fraction of 80cm)	Flash-over	Duration (min)	Ignition Time of Day (hr:min)
ISONG1	9/7/2007	Natural Gas	Pool Fed	1	1	1	N	37	12:41
ISONG2	9/10/2007	Natural Gas	Pool Fed	1	1	1	N	33	15:38
ISONG3	9/11/2007	Natural Gas	104	1	1	1	N	70	14:31
ISOHept4	9/12/2007	Heptane	Pool Fed	1	1	1	N	65	14:39
ISOHept5	9/13/2007	Heptane	Pool Fed	1	1	1	N	75	14:16
ISOHept8	2/27/2008	Heptane	10	1	0.5	0.25	Y	6	11:13
ISOHept9	2/28/2008	Heptane	20	1	0.5	0.25	Y	10	14:32
ISONylon10	2/29/2008	Nylon	10	1	0.5	0.25	N	30	10:35
ISOPP11	2/29/2008	PolyProp	10	1	0.5	0.25	N	35	14:22
ISOHeptD12	3/3/2008	Heptane	20	2	0.25	0.25	Y	10	11:32
ISOHeptD13	3/4/2008	Heptane	20	2	0.25	0.25	Y	10	10:15
ISOPropD14	3/4/2008	Propanol	24	2	0.25	0.25	Y	12	2:09
ISOProp15	3/5/2008	Propanol	24	1	0.5	0.25	Y	10	10:55
ISOStyrene16	3/5/2008	PolyStyrene	10	1	0.5	0.25	N	35	14:32
ISOStyrene17	3/7/2008	PolyStyrene	30	1	1	0.25	Y	37	11:53
ISOPP18	3/10/2008	PolyProp	20	2	0.5	0.25	Y	40	10:37
ISOHept19	3/10/2008	Heptane	20	1	0.5	0.25	Y	10	14:07
ISOToluene20	3/11/2008	Toluene	17	1	0.5	0.25	Y	10	10:56
ISOStyrene21	3/11/2008	PolyStyrene	15	1	0.5	0.25	N	35	14:47
ISOHept22	3/12/2008	Heptane	spray	1	0.5	0.25	Y	45	10:43
ISOHept23	3/12/2008	Heptane	spray	1	0.5	0.25	Y	40	3:10
ISOHept24	3/13/2008	Heptane	spray	1	0.5	0.125	Y	40	10:43
ISOHept25	3/13/2008	Heptane	spray	1	0.5	0.5	Y	20	14:23
ISOHept26	3/14/2008	Heptane	spray	1	0.5	0.5	Y	20	10:27
ISOHept27	3/17/2008	Heptane	spray	1	0.5	0.125	Y	40	10:48
ISOHept28	3/17/2008	Heptane	spray	1	0.5	0.25	Y	20	14:48
ISOToluene29	3/19/2008	Toluene	spray	1	0.5	0.25	Y	35	10:50
ISOPropanol30	3/19/2008	Propanol	spray	1	0.5	0.25	Y	35	1:40
ISOPUF31	3/20/2008	Polyurathane Foam	2.5	1	0.5	0.25	N	10	10:11
ISONG32	3/20/2008	Natural Gas		1	0.28	0.25	Y	25	12:01

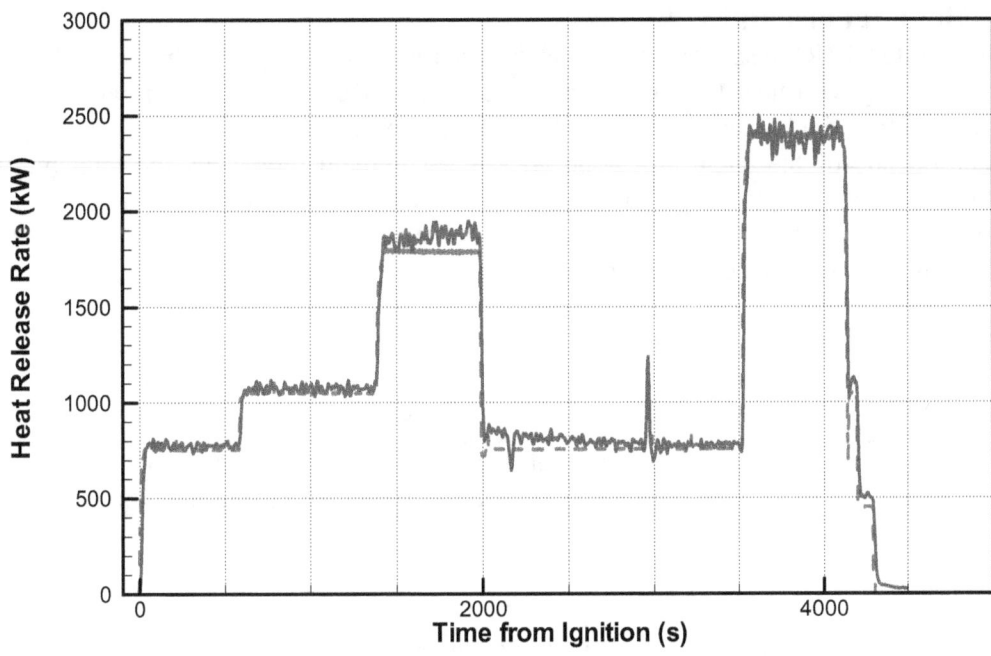

Figure 3.1: Heat release rate for test ISONG3 comparing the ideal heat release rate, as imposed by gas flow rate, by the red dashed line and the measured, by oxygen loss calorimetry, solid blue line.

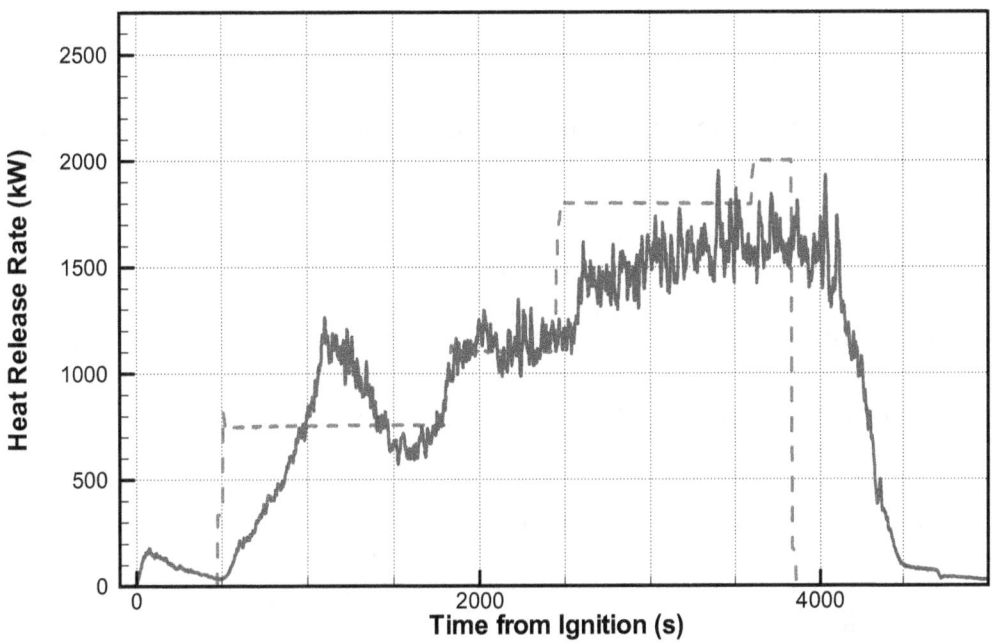

Figure 3.2: Heat release rate for test ISOHept5 comparing the ideal heat release rate, as imposed by pump flow rate, by the red dashed line and the measured, by oxygen loss calorimetry, solid blue line.

3.2 Heat Release Rate

The heat release rate (HRR) measurement was used to characterize the size of the fire and also to help determine (along with heat flux data, temperatures, and gas species concentration trends) when the fire conditions had reached steady state. As the fire becomes under-ventilated burning can take place outside of the enclosure. The HRR measurement represents the total burning inside and outside of the enclosure. Table 3.2 shows a description of the measurement labels used in the table column headings and figure legends in this section. These labels are identical to the column headings in the reduced data files.

Table 3.2: Description of calorimetry measurement labels.

Measurement Label	Description
HRR	Heat Release Rate from Calorimeter, kW
IHRR	Ideal Heat Release Rate from Burner (gas, pool or spray), kW

The heat release rate of a natural gas fueled fire is shown in Figure 3.1. The flow of natural gas was precisely metered with a gas flow valve, and the surface area in this test was controlled by the surface are of the gravel in the burner which did not change. The gaseous fed burner was much easier to control in the fire than the liquid fuel burners because there was no thermal induced time lag between fuel delivery rate and the burning rate. In contrast, the heat release rate results for a heptanes fueled fire test are shown in Figure 3.2. In that case the fire was never steady due to instability in the burning rate process control variables. Maintaining a constant burning rate was difficult because of the coupled effects of fuel cooling, due to the water cooled burner, natural burning quickly changing the surface area of the fuel and a manually controlled fuel delivery system.

Figure 3.3 shows the heat release rate from a free burn of heptanes in test ISOHept9. In this test, a fixed quantity of fuel, 30 L, was placed in a constant surface area, 0.5 m^2 pan burner, c.f. Figure 2.4, located in the center of the floor, Position 1, c.f. Figure 2.5. A 20 cm (1/4 width) doorway was used. The fuel was ignited and the mass loss from the pan was measured simultaneously to the oxygen calorimetry, both are shown in Figure 3.3. The difference between the measured mass loss, ideal HRR (IHRR), and the measured calorimetry is related to the combustion efficiency of the fire. The ideal HRR leads the dynamic behavior of the measured HRR and there is a proportional response of the two values. The HRR values from this type of free burn are not extremely steady; they fluctuate due to the dynamic and highly turbulent nature of the fire. However, the fire was steadier than ISOHept5, described in Figure 3.2, and more accurately simulated a real situation than ISONG3, presented in Figure 3.1. Generally speaking, the fire was ignited and then the HRR ramps up over a period of about 70 s to approach a pseudo-stead state burning of the fuel in the pan which was maintained until the fuel evaporated from the pan. The ideal heat release rate preceded the measured heat release rate because the fuel vaporized and ignited before any oxygen depletion was measured. Likewise, near the end of the test all of the fuel is gone from the pan, but still in vapor form for a short time leading to the ideal HRR dropping off sooner than the measured HRR.

Figure 3.4 presents the ideal and measured HRR of an iso-propanol fire in an identical configuration to that illustrated above for a heptanes fire, ISOHept9. In this case 30 L of propanol was used again with a 20 cm (1/4 width) doorway. Again, the fire was ignited and then approached a pseudo-steady burning rate. The differences in this case are that: 1) the iso-propanol fire took nearly 160 s to reach pseudo-steady state as compared to 70 s for the heptanes fire, 2) the test lasted nearly 700 s for iso-propanol while it only lasted 600 s for heptanes, and 3) the steady state heat release rate for iso-propanol was approximately 1200 kW while it was 2000 kW for the heptanes case. These differences can be partly attributed to the lower heat of combustion of iso-propanol when compared with heptanes. (Iso-propanol has a net heat of combustion of 30.45 MJ/kg and heptane has a neat heat of combustion of 44.4 MJ/kg [3].)

Figure 3.5 and Figure 3.6 present ISOHeptD12 and ISOHeptD13, two identical distributed heptanes fueled fire cases. Both cases were conducted with two 0.25 m^2 burners located in the center of the floor and against the back wall centered in the x-direction, positions 1 and 2, cf. Figure 2.5. Both tests were conducted with a 20 cm (1/4 width) doorway. There are two things to note here. First, these tests were conducted in an identical manner so that the repeatability of the experiment was established. Comparing Figure 3.5 and Figure 3.6 there were only minor differences in the dynamic behavior of the fires, and the global behavior of the two experiments was almost exactly the same. Second, the behavior of the distributed fuel sources relative to that of the single fuel source was examined. Both the quantity of the fuel and the surface area of the fuel in ISOHept9 were maintained globally. The fire was ignited and slightly overshot the steady state HRR before reaching the steady burning condition. The steady burning condition of the distributed fuel case had a similar heat release rate to that observed for the single burner case presented in Figure 3.3. Near the end of the test there was a sharp drop off in both the ideal and measured heat release rates prior to the fuel burn out. A second steady burning rate was established for approximately 100 s. This drop off in HRR was due to the fuel in the rear burner being consumed faster than the fuel in the front burner. This was confirmed by observing that the mass of fuel measured in the rear burner reached zero while the middle burner still showed some fuel mass prior to the fire going out. An attempt was made to verify this visually, however due to the intense nature of the fire and the narrow door opening it was not possible, cf. Figure 3.7.

Figure 3.8 presents the heat release measurements from test ISOStyrene17, a burn with 30 kg of polystyrene on a 1 m^2 burner situated in the center of the room, position 1 cf. Figure 2.5, and a 20 cm (1/4 width) doorway. The global intent of this study was to investigate under-ventilated fires and previous tests, e.g. ISOStyrene16, with a smaller surface area of polystyrene failed to provide under-ventilated conditions. This was the only test done with the 1 m^2 pan burner because the size of the burner caused it to warp significantly during the test. Figure 3.9 shows images of the pan warping and the front edge of the pan lifting off of the enclosure floor by as much as 20 cm during the test. This resulted in a failure of the mass loss measurement, Figure 3.8, because the burner was in contact with the enclosure floor. A similar failure can be seen in Figure 3.10, ISOPPD18, where polypropylene was burned in two 0.5 m^2 pans at the center and center rear of the enclosure, positions 1 and 2 cf. Figure 2.5. In both cases the ideal HRR follows the measured HRR near the beginning and end of the test, however in the middle of the test, specifically during flashover the burners were warped due to excessive temperature. Excessive burner temperatures are more likely to occur with solid fuels because vaporization of the fuel happens at a much higher temperature than it does for liquid fuels, thus the burner will likely experience a higher temperature even if more heat is absorbed by the fuel.

Figure 3.11 presents the heat release rate results for the heptanes spray burner case, ISOHept22. The spray nozzle was positioned 25 cm above the floor in the center of the room. The 0.5 m^2 pan burner was used in position 1, cf. Figure 2.5 and allowed for measurement of any fuel collecting in the pan. The spray burner allowed for much better control over the heat release rate in the room than either the fed pool or the natural burning. Utilizing a spray burner was not as realistic as freely burning fuel, however there was a strong advantage to having a very steady HRR for a period of minutes for validation of numerical simulations. The spray burner experiments allow for steady state comparisons of other features of the enclosure, such as scaling of the doorway. An extensive discussion of scaling of the doorway is presented in section 6.3.

Figure 3.12 presents the test, ISOHept27, where a HRR ramp created by linearly increasing the fuel delivery rate to the spray burner over a period of time. The room configuration was identical to that discussed above for ISOHept22. The ideal HRR in Figure 3.12 increased faster than the measured HRR, as expected. As more fuel was pumped into the room there was, relatively speaking, less oxygen present for the fuel to consume. A combination of the HRR data with the oxygen species in the room allows determination of when the under-ventilated conditions were achieved. A thorough discussion of the HRR ramp test is presented in section 6.4.

A summary of steady state heat release rates for many of the fuels investigated here is presented in Figure 3.13. The dashed line indicates ideal or complete burning of the fuel in each case. The ideal burning rate was determined from either the mass loss rate (derived from load cell mass data), the liquid fuel flowrate, or the gaseous fuel flow rate as appropriate for each test. As expected and in agreement with Ref. [5] the combustion efficiency of the cleaner burning fuels (e.g. natural gas) was closer to ideal than the highly sooting fuels (e.g. toluene). For most fuels, the global combustion efficiency decreased as the fire became more under-ventilated. A more thorough discussion of combustion efficiency can be found later in section 4.3. A tabulated summary of the averaged steady-state results used to produce Figure 3.13 are presented in Table 3.3.

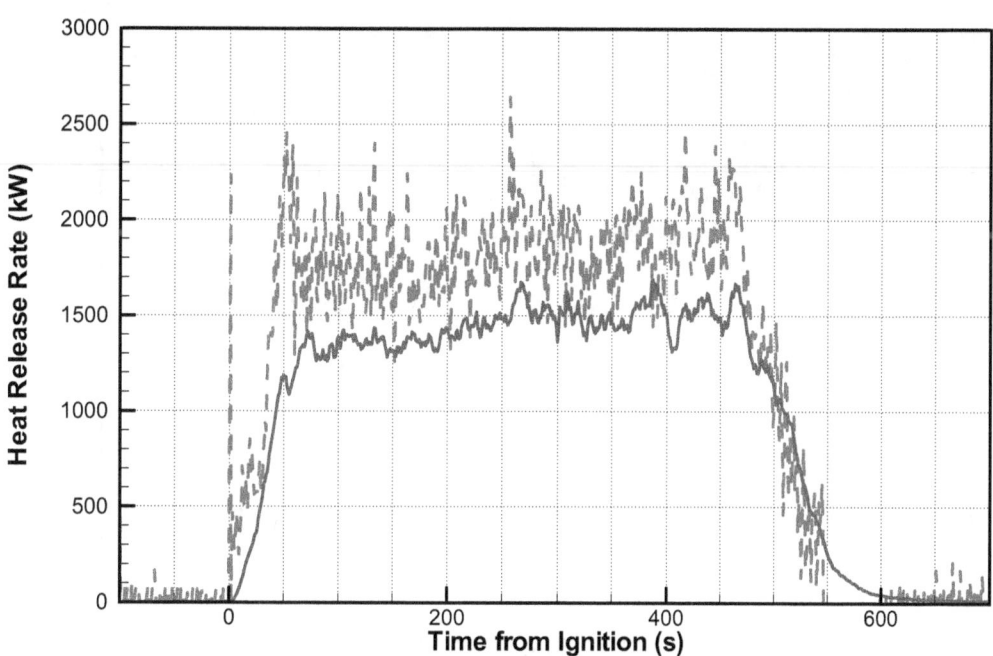

Figure 3.3: Heat release rate for test ISOHept9 (Heptane) comparing the ideal heat release rate, as measured by the burner mass loss rate, by the red dashed line and the measured, by oxygen loss calorimetry, solid blue line.

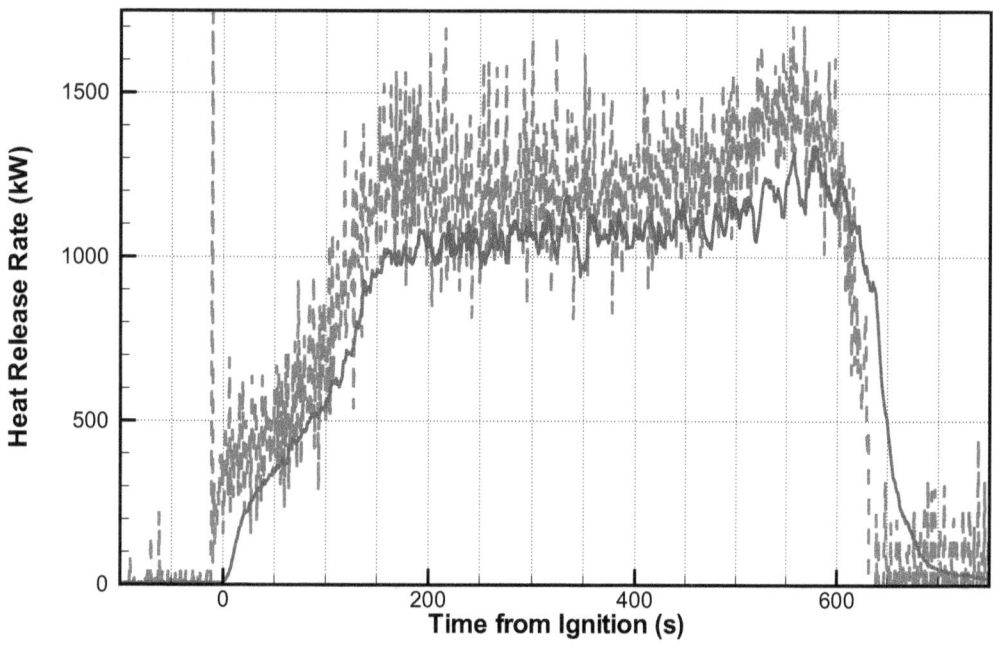

Figure 3.4: Heat release rate for test ISOProp15 (Iso-Propanol) comparing the ideal heat release rate, as measured by the burner mass loss rate, by the red dashed line and the measured, by oxygen loss calorimetry, solid blue line.

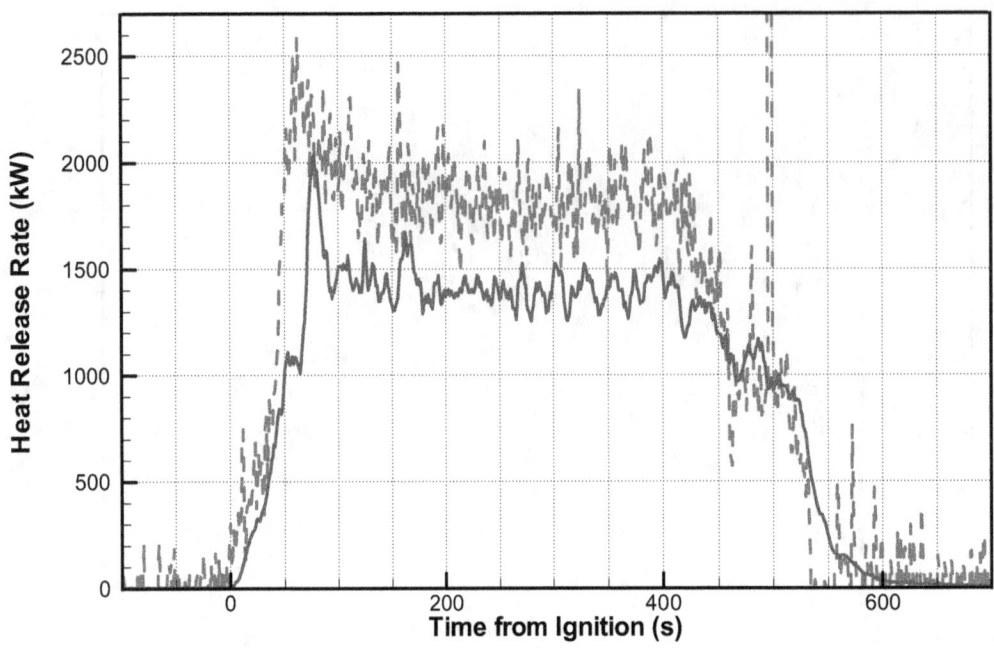

Figure 3.5: Heat release rate for test ISOHeptD12 (Heptane) comparing the ideal heat release rate, as measured by the burner mass loss rate, by the red dashed line and the measured, by oxygen loss calorimetry, solid blue line.

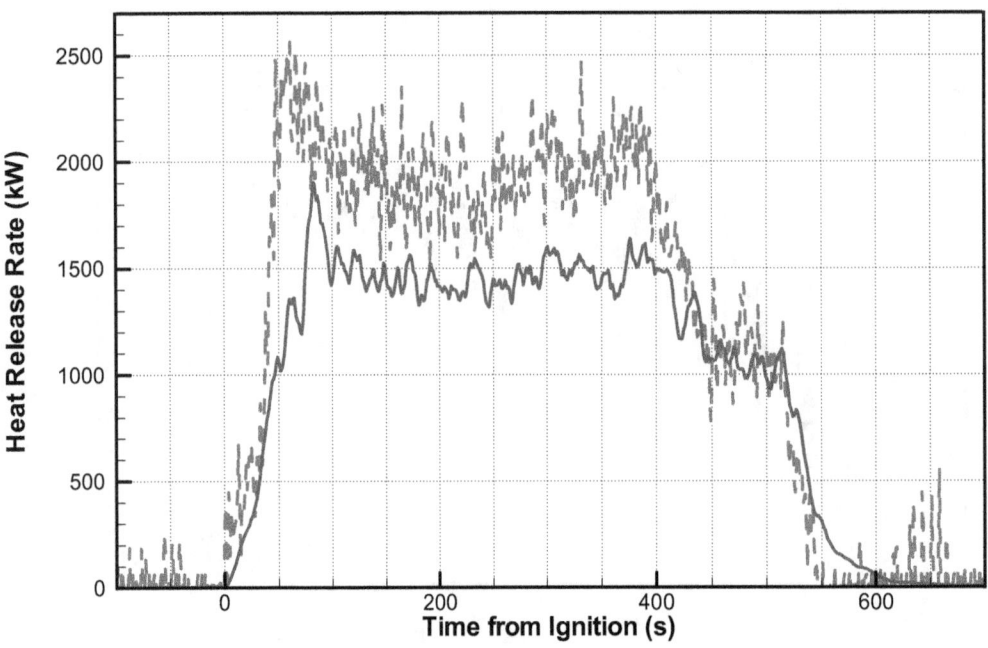

Figure 3.6: Heat release rate for test ISOHeptD13 (Heptane) comparing the ideal heat release rate, as measured by the burner mass loss rate, by the red dashed line and the measured, by oxygen loss calorimetry, solid blue line.

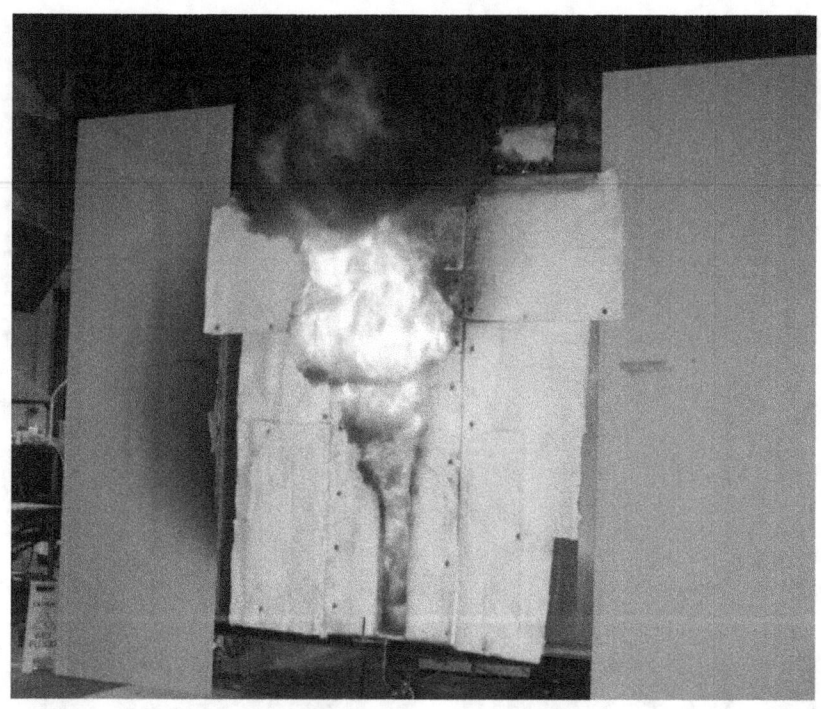

Figure 3.7: Fire leaving the door of the ISO 9705 room during test ISOHeptD12. It was not possible to view the inside of the room during this test.

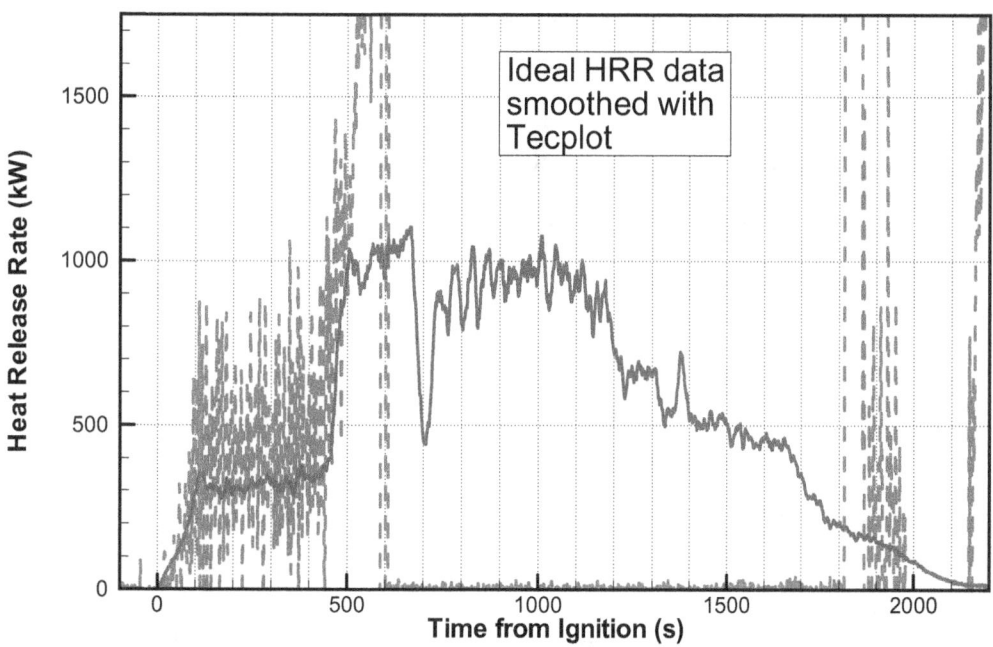

Figure 3.8: Heat release rate for test ISOStyrene17 (Polystyrene) comparing the ideal heat release rate, as measured by the burner mass loss rate, by the red dashed line and the measured, by oxygen loss calorimetry, solid blue line. The mass loss reading was lost during the experiment due to warping of the burner pan. The ideal HRR values were smoothed because of excessive signal noise.

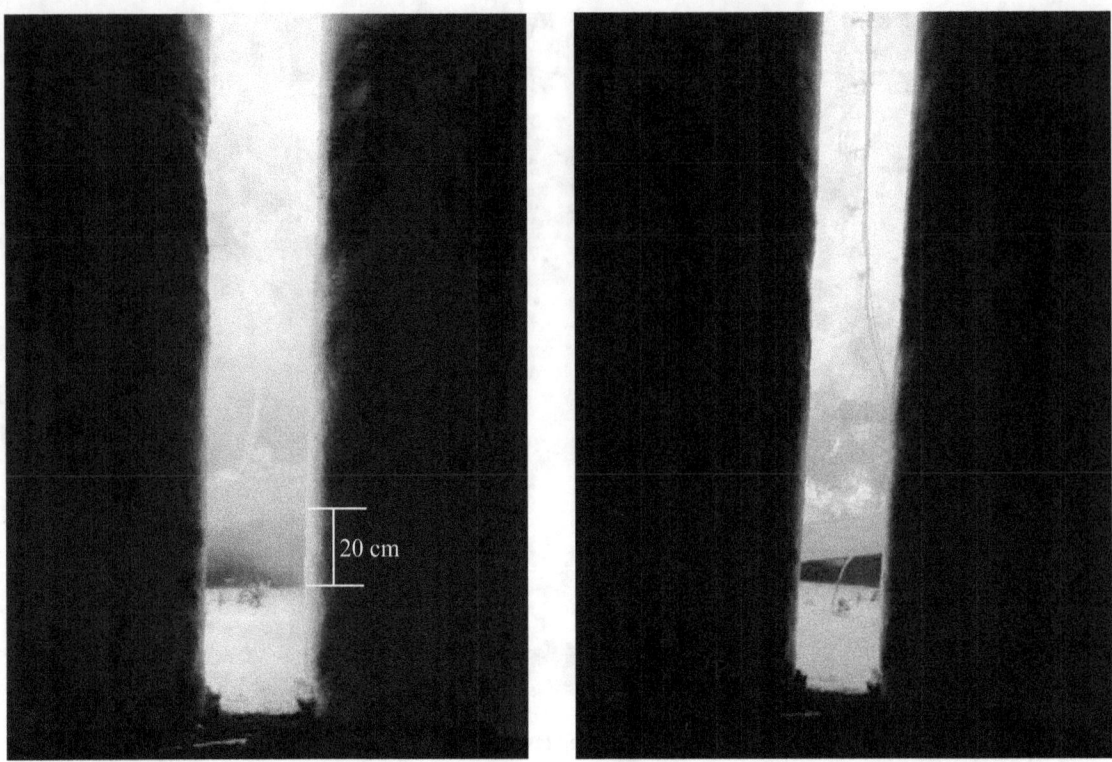

Figure 3.9: Images of the burner warping and moving during test ISOStyrene17. The burner was observed to be as much as 20 cm off of the floor in one corner. The burner warped because of its large size, 1m^2, and the excessive heat transfer to the burner in the room. This caused a loss of the mass loss measurement in this experiment.

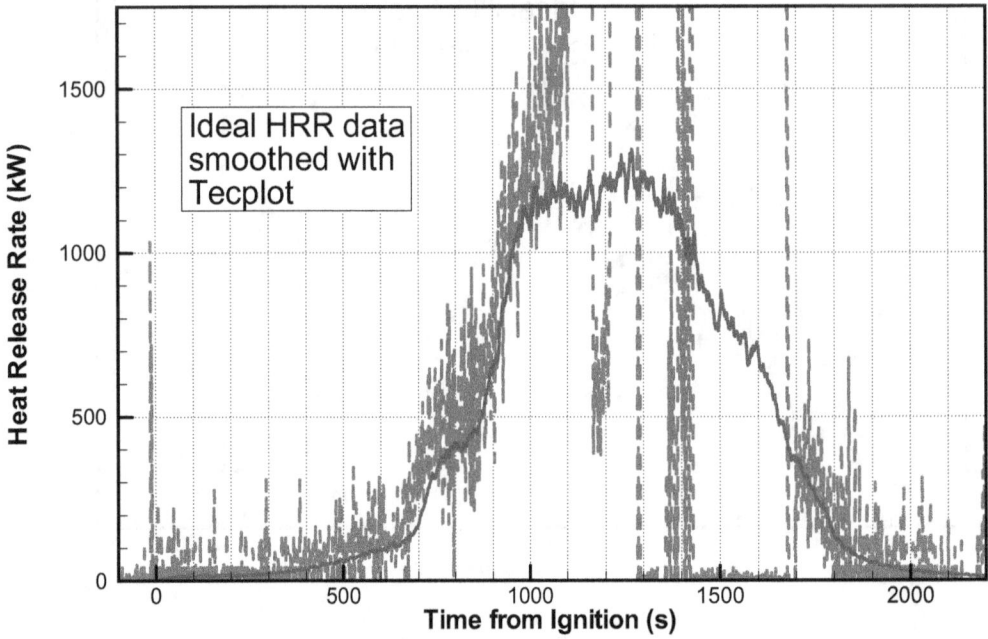

Figure 3.10: Heat release rate for test ISOPPD18 (Polypropylene) comparing the ideal heat release rate, as measured by the burner mass loss rate, by the red dashed line and the measured, by oxygen loss calorimetry, solid blue line. The mass loss reading was lost during the experiment due to warping of the burner pan. The ideal HRR values were smoothed because of excessive signal noise.

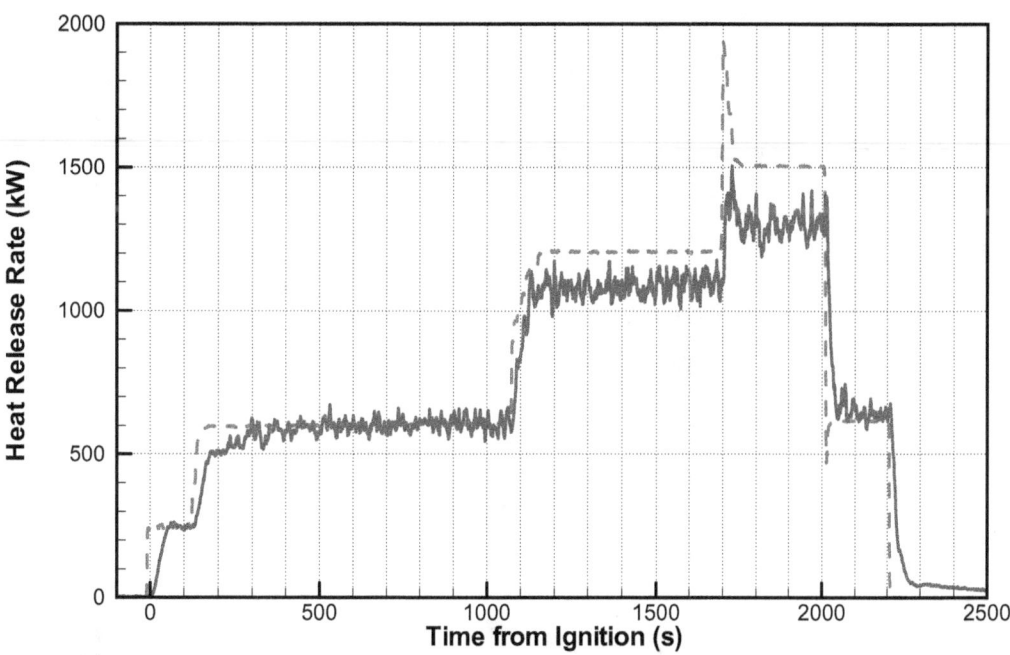

Figure 3.11: Heat release rate for test ISOHept22 (heptane) comparing the ideal heat release rate, as measured by the spray burner pump flow rate, by the red dashed line and the measured, by oxygen loss calorimetry, solid blue line.

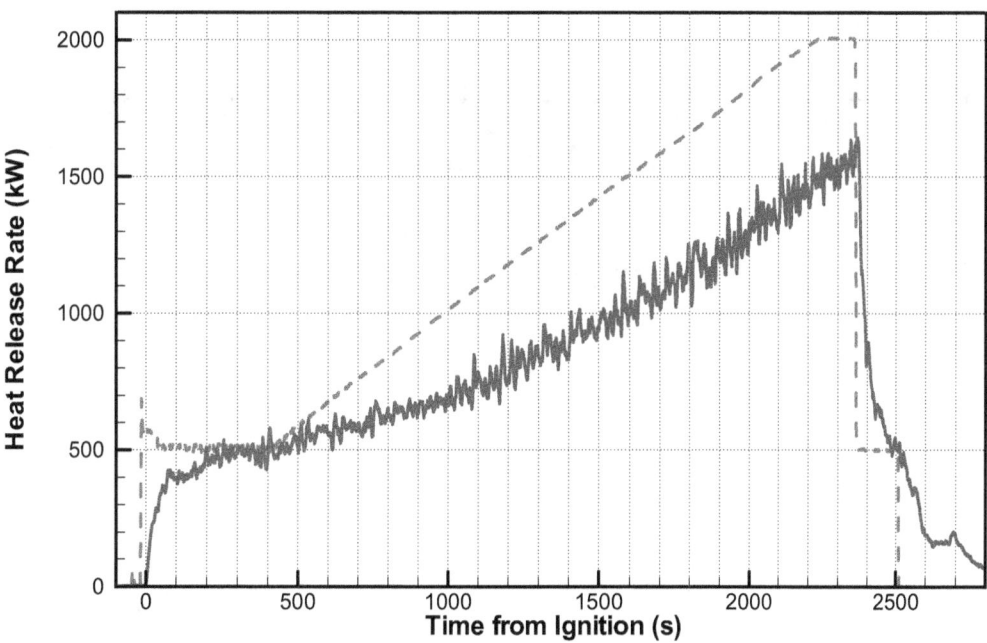

Figure 3.12: Heat release rate for test ISOHept27 (heptane) comparing the ideal heat release rate, as measured by the spray burner pump flow rate, by the red dashed line and the measured, by oxygen loss calorimetry, solid blue line. This test featured a linear increase in the fuel delivery rate to observe the effects of a HRR ramp.

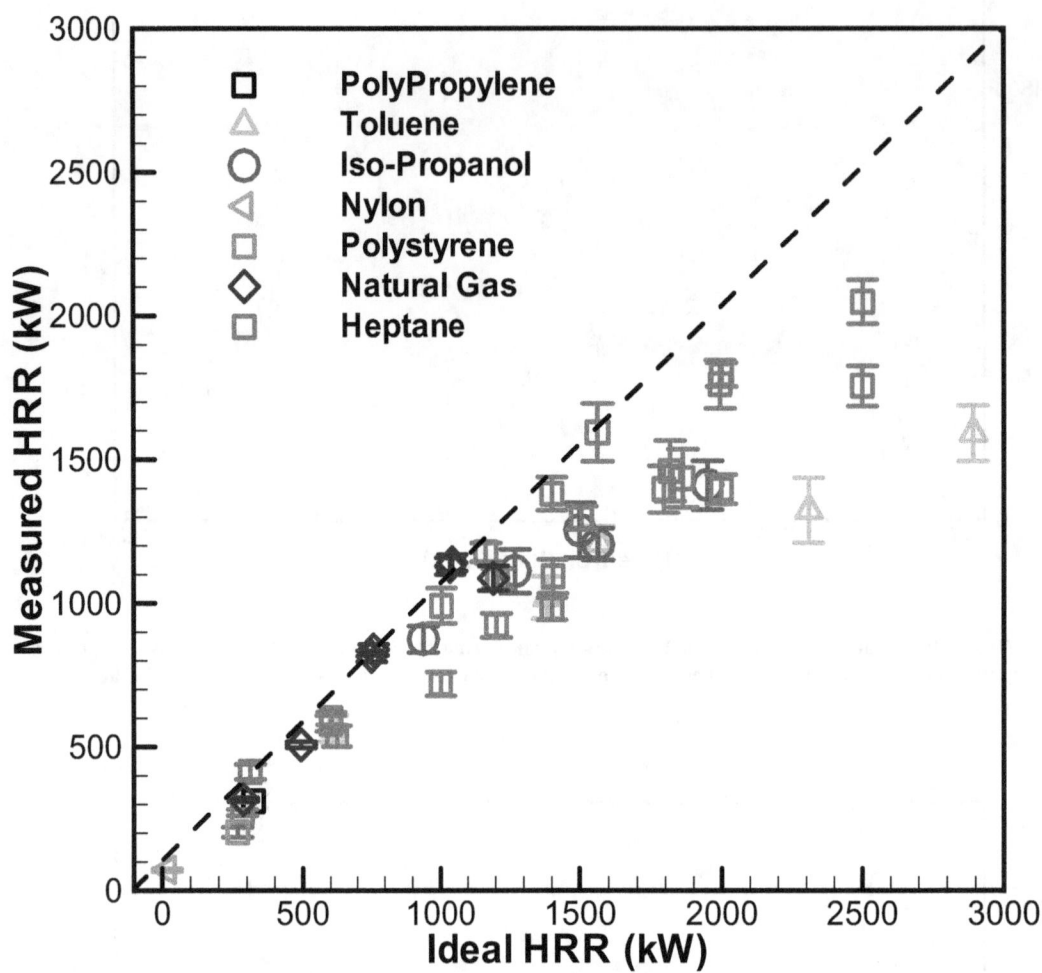

Figure 3.13: steady state heat release results. The dashed line indicates ideal or complete burning.

Table 3.3: Summary of averaged steady-state results of HRR and exhaust stack species measurements. *U* indicates the standard deviation in each steady state measurement.

Test No.	Fuel	Steady State Window		HRR cal		HRR ideal	
		start (s)	*stop (s)*	*Mean*	*U*	*Mean*	*U*
1	Natural Gas	350	800	295.38	6.0	287.7	2.8
		1300	1550	761.02	18.5	752.1	5.6
		1800	2100	1052.78	26.8	1034.8	2.7
2	Natural Gas	200	600	781.76	18.3	759.1	0.9
		800	1300	1063.78	26.4	1042.4	4.0
3	Natural Gas	700	1350	1074.84	27.5	1050.9	3.0
		1500	1900	1866.63	51.0	1789.6	5.5
		3650	4050	2382.37	76.6	2390.3	8.1
4	Heptane	3300	3700	2069.25	184.8	2403.1	1.13
5	Heptane	3100	3800	1490.86	98.0	1850.2	88.24
8	Heptane	170	230	1183.0	29.0	1164.6	354.9
9	Heptane	150	500	1460.2.4	251.1	1816.8	315.6
10	Nylon	650	1100	70.6	6.1	18.1	423.2
11	PolyPropylene	1100	1900	312.1	24.0	323.6	283.9
12	Heptane	125	450	1401.0	78.2	1794.0	237.0
13	Heptane	125	450	1448.4	81.7	1861.9	309.7
14	Iso-Propanol	280	460	1239.6	68.4	1500.3	149.9
16	Polystyrene	800	1600	204.5	17.7	267.4	196.7
15	Iso-Propanol	250	600	1107.2	75.8	1268.7	174.0
17	Polystyrene	800	1175	0.0	0.0	0.0	38.7
19	Heptane	125	425	1510.0	99.7	N/A	N/A
18	PolyPropylene	1100	1425	1181.4	53.3	N/A	N/A
21	Polystyrene	950	1400	696.7	36.2	N/A	N/A
20	Toluene	150	450	1207.4	59.0	1566.9	311.9
22	Heptane	350	1050	600.1	22.3	603.6	3.3
		1250	1700	1085.5	34.4	1209.2	38.4
		1775	2000	1297.1	41.7	1506.0	1.4
23	Heptane	125	500	270.9	10.9	284.4	1.9
		725	975	538.4	36.1	629.6	42.3
		1050	1350	719.0	40.8	1001.1	1.1
		1475	1725	980.7	38.1	1400.8	1.9
24	Heptane	200	360	414.0	26.2	321.8	55.8
		450	850	581.6	27.3	601.0	1.0
		950	1450	923.4	41.5	1202.7	1.5
		1650	1950	1092.7	59.4	1404.4	1.2
25	Heptane	350	575	991.0	61.5	1003.3	0.9
		650	1000	1380.9	57.6	1403.4	18.3
		1050	1350	1762.2	84.4	1993.5	3.6
26	Heptane	325	575	1795.6	41.6	1998.8	0.9
		640	800	2048.5	76.3	2527.6	129.3
28	Heptane	250	650	1396.6	50.9	1997.9	2.7
		750	1100	1756.0	70.6	2500.1	1.5
30	Iso-Propanol	400	800	875.3	46.2	937.4	0.6
		1350	1600	1204.7	55.3	1563.2	1.0
		1650	1850	1411.2	84.8	1948.9	55.8
29	Toluene	180	320	564.1	46.0	623.7	104.5
		400	700	1020.2	73.2	1383.7	1.3
		800	1150	1323.3	113.0	2309.0	1.6
		1600	1760	1591.2	96.6	2890.0	32.4
32	Natural Gas	200	300	507.4	10.7	494.4	0.4
		500	1000	1086.6	42.9	1192.3	6.8

3.3 Temperatures

When a bare bead thermocouple is measuring a high temperature but is exposed to a relatively cool ambient the thermocouple looses heat to that ambient due to thermal radiation and reports an incorrect, lower, temperature. This is a problem when trying to instrument experiments which incorporate high temperature gases such as jet engines or fires. To combat this problem an aspirated thermocouple was developed [34] which includes a double radiation shield and pulls hot gases over the shields and thermocouple bead at a high rate, as described in section 2.2.5.1. However, the utilization of aspirated thermocouple requires the sacrifices of temporal and spatial resolution. In addition, this is constraining when many thermocouples are used simultaneously in a compartment fire since a vacuum flow for each aspirated TC is needed. Thus, in this study, bare bead thermocouples were primarily used to measure temperature. In order to validate the results aspirated and bare bead thermocouples were compared side by side in tests ISONG1, ISONG2, ISONG3, ISOHept4, and ISOHept5.

Figure 3.14 shows a comparison between temperatures measured using bare bead and aspirated thermocouples at front sampling location as a function of time for natural gas test ISONG3. The heat release rate measured from calorimeter was also plotted to show fire conditions. Refer to Table 2.5 and Figure 2.1 for exact locations of the temperature probes. The measurement labels used in the table column heading and figure legends in this section are described in Table 3.4. For all locations, bare bead thermocouples provide lower temperatures compared to those of aspirated thermocouples due to the effect of radiative losses. At the front sample location, bare bead (TFSampPtRh) and aspirated thermocouples (TFSampA) show similar time histories of temperature qualitatively. However, the difference between these thermocouples decreases for heat release rates greater than approximately 1200 kW. At the rear sample location, the difference between bare bead and aspirated thermocouple temperatures is smaller compared to those at the front sample location.

Figure 3.15 presents a comparison of the averaged bare bead and aspirated thermocouple temperatures for the same natural gas test, ISONG3, shown in Figure 3.14. Averaged temperatures were calculated over pseudo-steady periods and averaged periods are described in the figure. The difference between the two thermocouples decreases as the heat release rate increases. For example, at the front sample location, the relative difference of bear bead and aspirated thermocouple measurements decreases from 4.06 % to 0.83 % when heat release rate increases from 1203 kW to 2667 kW. These variations correspond to the temperature differentials of 37 °C and 10 °C, respectively. The decrease in the difference between the aspirated and bare bead temperatures at larger fire size is likely due to larger amounts of soot being produced in the room and blocking radiative losses from the bare bead thermocouple. In addition, the rear sample location, better agreement between bear bead and aspirated thermocouple measurements is observed compared to the measurements at the front sample location. This may be expected because the front thermocouple location is likely to have greater radiative heat exchange with a cool ambient via the doorway. These results illustrate the performance of bare bead thermocouples in this experimental configuration and provide confidence that the bare bead thermocouples are providing accurate results.

To demonstrate the reproducibility of the measurements over a number of different tests, Figure 3.16 shows averaged temperatures measured at the front and rear thermocouple trees as a function of the height above the floor for ISOHetpD12 and ISOHeptD13. These tests were two identical distributed heptanes fueled fire cases and conducted one day apart so that the reproducibility of the experiment could be established. As mentioned earlier, there were only minor differences in the dynamic behavior of the fires in terms of heat release rate. In this figure, the vertical profiles of temperature at the front location for the two cases are nearly identical. The temperature profiles at the rear location also show only minor differences between the two cases except for the measurements below 0.9 m. From this figure it is clear that temperature measurements are reproduced well between ISOHeptD12 and ISOHeptD13.

Table 3.4. Description of interior gas temperature measurement labels.

Measurement Label	Description
TFSampPtRh	Bear bead thermocouple at front sample location (208 cm above floor)
TRSampPtRh	Bear bead thermocouple at rear sample location (208 cm above floor)
TFSampA	Aspirated thermocouple at front sample location (208 cm above floor)
TRSampA	Aspirated thermocouple at rear sample location (208 cm above floor)
TF3	Bear bead thermocouple at front location (2.5 cm above floor)
TF30	Bear bead thermocouple at front location (30 cm above floor)
TF60	Bear bead thermocouple at front location (60 cm above floor)
TF180	Bear bead thermocouple at front location (180 cm above floor)
TF210	Bear bead thermocouple at front location (210 cm above floor)
TF237	Bear bead thermocouple at front location (237.5 cm above floor)
TR3	Bear bead thermocouple at rear location (2.5 cm above floor)
TR30	Bear bead thermocouple at rear location (30 cm above floor)
TR60	Bear bead thermocouple at rear location (60 cm above floor)
TR180	Bear bead thermocouple at rear location (180 cm above floor)
TR210	Bear bead thermocouple at rear location (210 cm above floor)
TR237	Bear bead thermocouple at rear location (237.5 cm above floor)

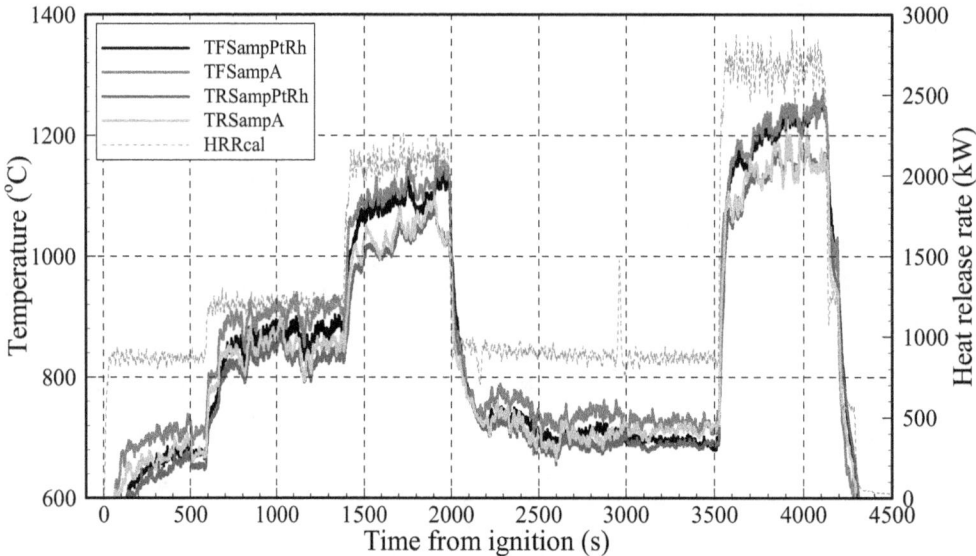

Figure 3.14: Comparison between temperatures measured from bare bead and aspirated thermocouples at front sampling location as a function of time for test ISONG3.

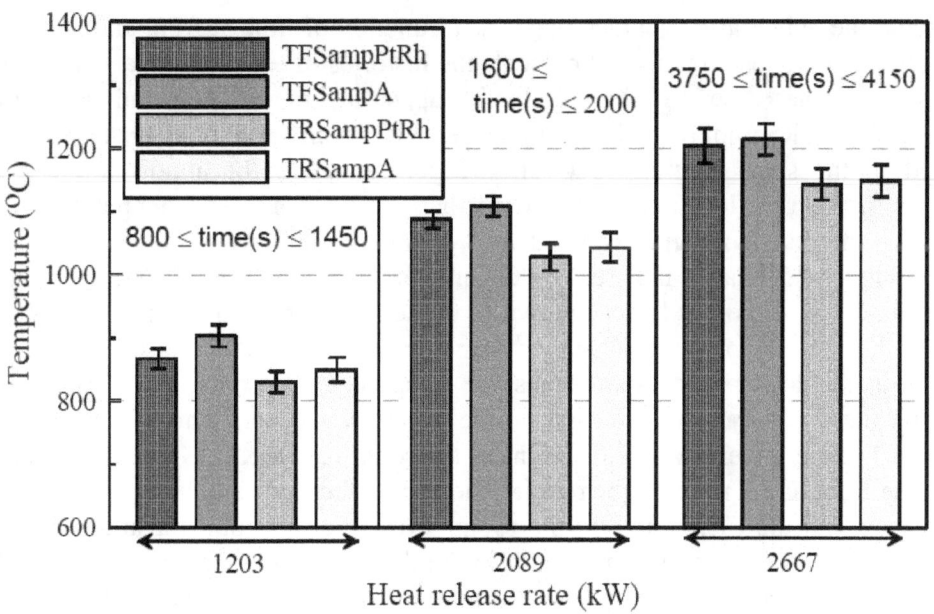

Figure 3.15: Comparison of averaged temperatures measured from bare bead and aspirated thermocouples at front sample location for test ISONG3.

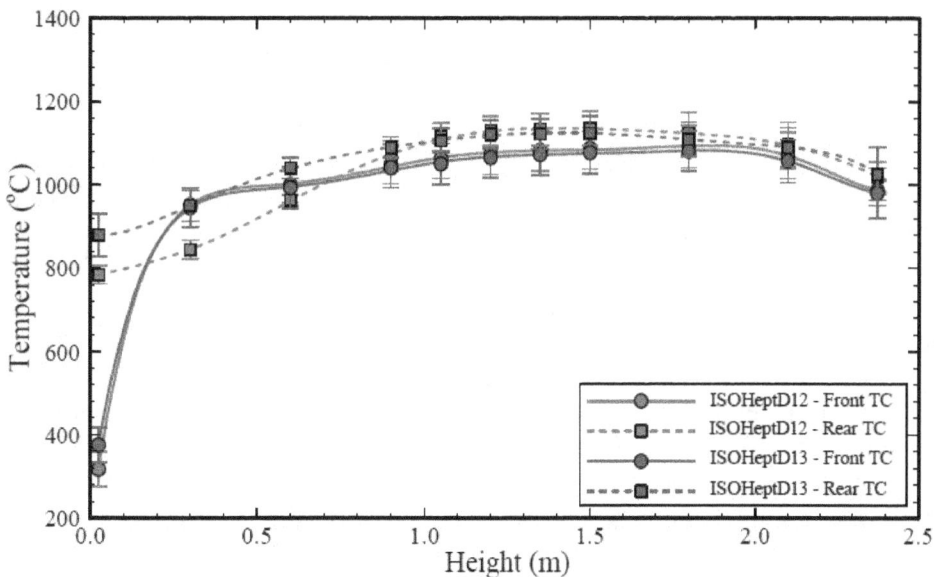

Figure 3.16: Comparisons of averaged temperature measured at front and rear thermocouple trees for test ISOHeptD12 and ISOHeptD13.

To understand the temperature characteristics as a function of time at front and rear locations of the upper and lower layer, Figure 3.17 presents the histories of temperature at 4 locations for the heptanes spray burner case, ISOHept22. In the upper layer, the front and rear temperatures (TFSampPtRh and TRSampPtRh) show similar behaviors compared to the change of heat release rate plotted in the same figure. However, the general trend for almost all tests is higher temperatures at the rear location in the room than at the front. The temperature differences between front and rear location are attributed to ventilation effects in the fire. It was observed here and previously [5] that as the area of the vent is reduced in an enclosure the temperature in the front of the room becomes lower relative to the temperature at the rear of the room. In the lower layer the rear temperature is higher than the front temperature. In particular, the rear temperature in the lower layer approaches the upper layer temperature. This result may be explained by the fire's dynamic flow inside the room. As cool air enters the under-ventilated compartment fire the lower level of the front of the room is cooled. As the air moves along the floor from the front of the room to the rear it is heated by the floor and hot gases in the room and begins to react with the fuel source in the room. Rear thermocouples also have smaller view angles of the cool external environment outside the door so they are more effectively heated by the thermal radiation from the upper layer and interior room surfaces, especially when high soot levels make the room gases optically thick.

Figure 3.18 and Figure 3.19 show the histories of temperature at front and rear thermocouple trees, respectively, from a free burn of heptanes in test ISOHept9. The front temperatures as a function of height show that temperatures increase gradually with height and then decrease again above 180 cm. There is a difference in temperature of 900 °C between 2.5 cm and and the maximum temperature at 180 cm. At the rear location as presented in Figure 3.19, temperatures located above 60 cm show a similar qualitative behavior to that observed in the front of the room. This indicates that the top of the upper layer in the room (180 cm and higher) is very uniform thermally. The temperatures at 2.5 cm and 30 cm decrease with time while the higher locations increase in temperature with time. This is part of the initial transient of the room where the initial fire, prior to becoming under-ventilated, would have been burning closer to the burner thus producing more heat near the floor. Another contribution to this effect is that as the fire becomes under-ventilated, most of the burning (and heat release) occurs near the front. The rear is deprived of oxygen and the actual flame sheet is confined to the front of the enclosure. The hot products are still convected upward and heat the whole upper layer, but rear surfaces, including the floor, receive less radiation from the flame sheet. Similar results are observed in other tests, such as ISOPropD14, ISOProp15, ISOStyrrene17, ISOHept23 and ISOHept25. Note that the rear thermocouple at 60 cm was about 100 °C to 150 °C higher than the front thermocouple at the same height. This indicates a very hot region closer to the floor in the rear possibly near the layer interface where the flame sheet is located.

Figure 3.20 and Figure 3.21 present steady temperatures at the front and rear sample locations for all fuels tested in this study. In general, the soot producing fires (heptane, polypropylene, polystyrene, toluene) produce hotter gas temperature inside the enclosure than the cleaner fires (natural gas) at the same measured heat release rate. These results also found in the previous experiments with the reduced-scale ventilation-limited compartment fires [5]. A summary of the averaged front and rear temperature measurements with combined expanded uncertainty (U) are listed in Table 3.5 and Table 3.6.

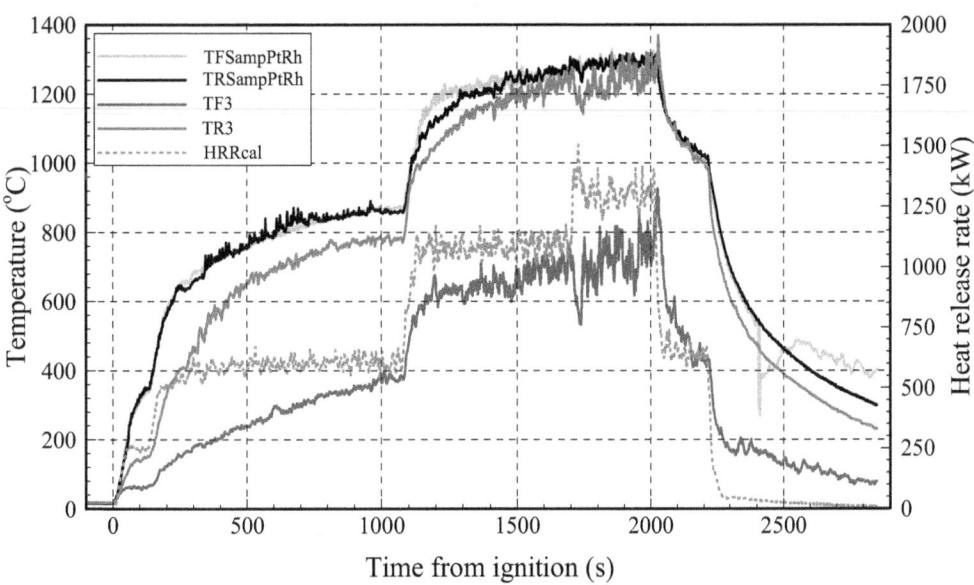

Figure 3.17: Histories of temperature at front and rear sampling locations for test ISOHept22 (heat release rate measured from calorimeter was included to show fire condition).

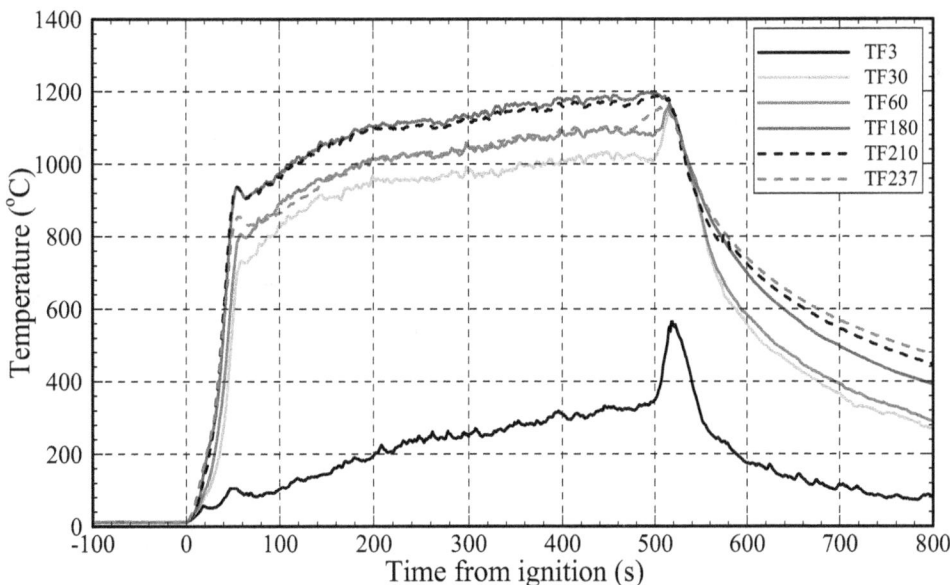

Figure 3.18: Histories of temperature at front thermocouple trees for test ISOHept9.

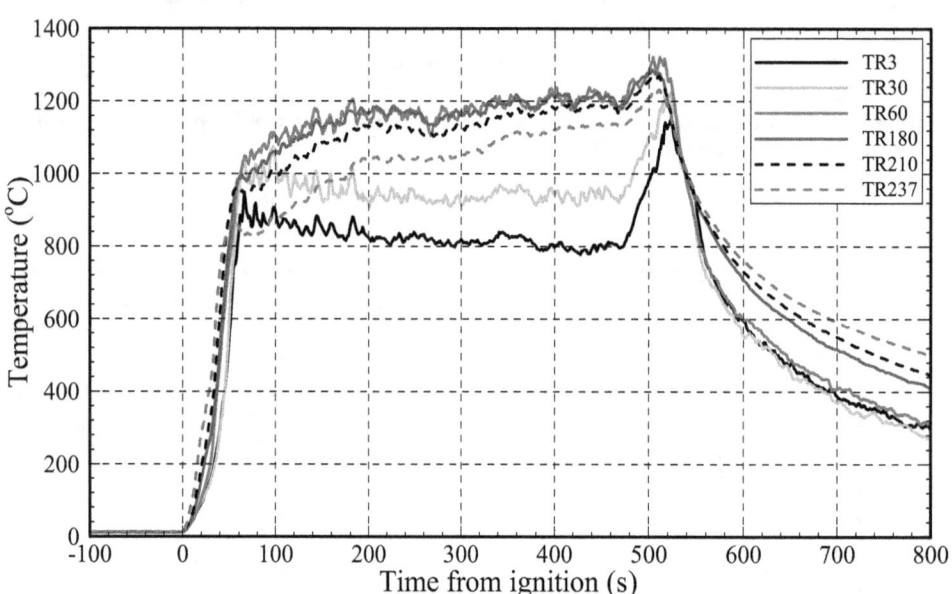

Figure 3.19: Histories of temperature at rear thermocouple trees for test ISOHept9.

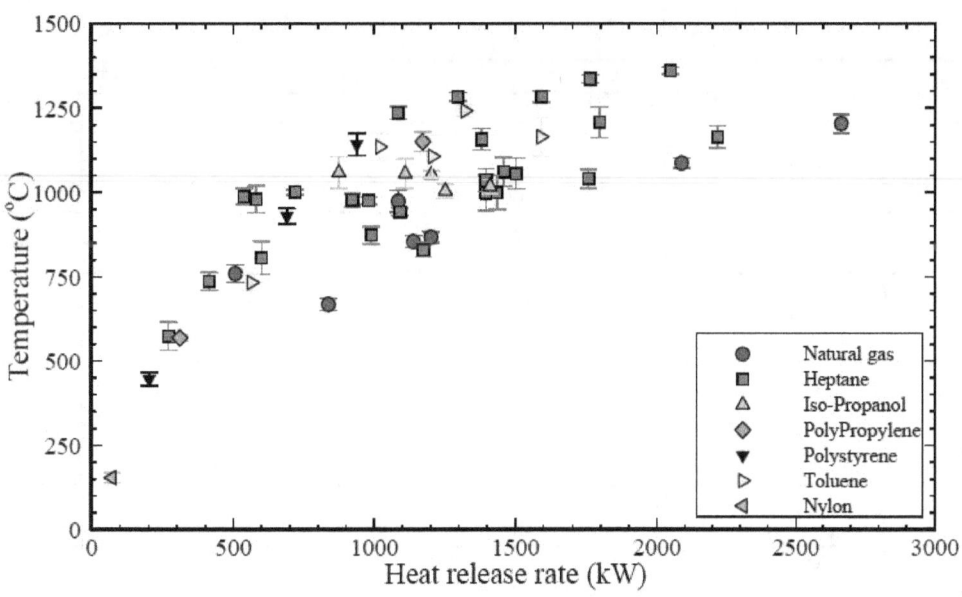

Figure 3.20: Averaged temperatures as a function of heat release rate at front sample location for all fuels tested.

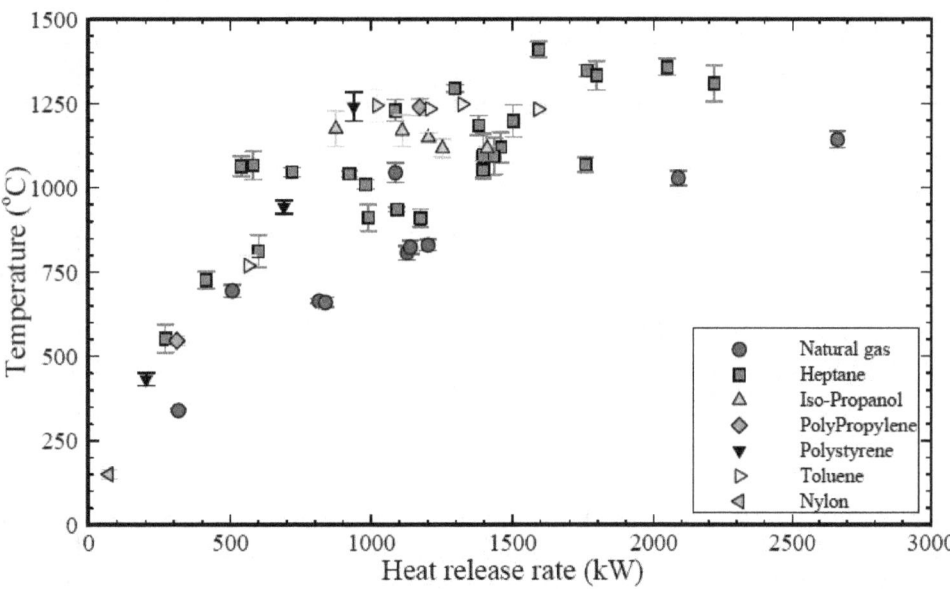

Figure 3.21: Averaged temperatures as a function of heat release rate at rear sample location for all fuels tested.

Table 3.5: Summary of averaged steady-state results of temperatures at front locations.

Test No.	Fuel	Steady State Window		TFSampA (°C)		TF3 (°C)	
		start (s)	stop (s)	Mean	U	Mean	U
1	Natural Gas	350	800				
		1300	1550				
		1800	2100				
2	Natural Gas	200	600	668	17.7		
		800	1300	854	18.4		
3	Natural Gas	700	1350	867	15.8		
		1500	1900	1087	14.1		
		3650	4050	1204	27.7		
4	Heptane	3300	3700	1164	32.9		
5	Heptane	3100	3800	1278.4	16.4		
8	Heptane	170	230	830	21.6	610.7	134.0
9	Heptane	150	500	1061	43.2	261.4	51.2
10	Nylon	650	1100	154	14.9	41.5	4.0
11	PolyPropylene	1100	1900	569	10.3	170.2	9.7
12	Heptane	125	450	999	53.6	317.9	41.3
13	Heptane	125	450	1000	51.0	376.0	42.3
14	Iso-Propanol	280	460	1005	20.9	350.9	35.4
16	Polystyrene	800	1600	446	19.8	79.9	3.7
15	Iso-Propanol	250	600	1056	43.3	326.4	43.5
17	Polystyrene	800	1175	1142	32.9	563.5	56.8
19	Heptane	125	425	1056	46.2	392.3	64.5
18	PolyPropylene	1100	1425	1151	29.4	472.6	106.4
21	Polystyrene	950	1400	929	23.3	404.5	29.8
20	Toluene	150	450	1107	61.7	483.9	76.2
22	Heptane	350	1050	806	48.9	295.8	55.5
		1250	1700	1235	19.0	662.5	35.6
		1775	2000	1283	13.1	725.3	50.7
23	Heptane	125	500	574	42.3	107.9	14.4
		725	975	989	24.6	360.5	32.5
		1050	1350	1001	8.0	453.0	42.8
		1475	1725	976	10.9	591.7	14.9
24	Heptane	200	360	736	25.9	236.7	23.2
		450	850	980	40.1	358.7	36.7
		950	1450	978	20.0	579.2	37.2
		1650	1950	942	11.9	711.5	28.5
25	Heptane	350	575	873	26.0	332.7	29.1
		650	1000	1158	31.6	645.8	21.3
		1050	1350	1336	11.6	904.5	55.1
26	Heptane	325	575	1207	45.2	613.8	88.8
		640	800	1360	11.5	847.0	39.6
28	Heptane	250	650	1037	34.4	444.0	60.4
		750	1100	1041	27.6	547.3	18.8
30	Iso-Propanol	400	800	1059	46.1	452.6	66.5
		1350	1600	1054	9.8	431.6	18.6
		1650	1850	1018	20.5	485.3	12.8
29	Toluene	180	320	733	24.0	171.6	21.0
		400	700	1135	41.3	426.0	76.3
		800	1150	1242	16.5	554.2	19.4
		1600	1760	1164	55.9	481.6	23.3
32	Natural Gas	200	300	759	25.6	176.2	14.8
		500	1000	974	32.5	646.2	69.2

Table 3.6: Summary of averaged steady-state results of temperatures at rear locations.

Test No.	Fuel	Steady State Window		TRSampA (°C)		TR3 (°C)	
		start (s)	stop (s)	Mean	U	Mean	U
1	Natural Gas	350	800	340	4.6		
		1300	1550	663	6.7		
		1800	2100	807	20.9		
2	Natural Gas	200	600	660	15.5		
		800	1300	823	20.1		
3	Natural Gas	700	1350	830	16.8		
		1500	1900	1028	21.5		
		3650	4050	1144	25.3		
4	Heptane	3300	3700	1309	53.3		
5	Heptane	3100	3800	1407.0	24.0		
8	Heptane	170	230	909	27.1	926.9	12.7
9	Heptane	150	500	1120	45.7	820.6	29.9
10	Nylon	650	1100	150	13.9	115.6	11.0
11	PolyPropylene	1100	1900	546	13.5	487.6	14.5
12	Heptane	125	450	1095	64.3	783.2	21.8
13	Heptane	125	450	1093	54.2	878.8	51.9
14	Iso-Propanol	280	460	1117	27.3	831.6	14.7
16	Polystyrene	800	1600	433	18.7	340.6	13.1
15	Iso-Propanol	250	600	1169	46.4	1047.8	126.0
17	Polystyrene	800	1175	1240	42.4	907.9	56.0
19	Heptane	125	425	1198	48.0	866.2	37.8
18	PolyPropylene	1100	1425	1239	25.2	980.8	96.8
21	Polystyrene	950	1400	943	19.5	954.9	28.9
20	Toluene	150	450	1234	67.3	1331.8	139.9
22	Heptane	350	1050	812	48.4	706.5	69.2
		1250	1700	1230	32.3	1180.8	41.5
		1775	2000	1294	10.3	1238.9	32.5
23	Heptane	125	500	553	41.9	450.7	57.5
		725	975	1063	28.9	1081.9	24.0
		1050	1350	1046	14.0	903.3	41.2
		1475	1725	1010	13.6	670.0	33.8
24	Heptane	200	360	727	25.8	659.2	31.8
		450	850	1066	42.1	931.2	51.5
		950	1450	1040	8.4	698.0	65.6
		1650	1950	936	6.7	582.7	6.3
25	Heptane	350	575	911	39.1	795.6	28.6
		650	1000	1185	29.5	1111.1	29.5
		1050	1350	1346	17.6	1266.1	58.3
26	Heptane	325	575	1333	43.0	1963.8	620.0
		640	800	1357	24.5	1381.9	1025.3
28	Heptane	250	650	1053	27.1	951.9	21.5
		750	1100	1068	21.6	923.2	10.9
30	Iso-Propanol	400	800	1175	52.6	1230.9	53.5
		1350	1600	1149	13.6	1015.1	19.8
		1650	1850	1117	5.7	954.3	13.7
29	Toluene	180	320	769	30.4	801.8	75.4
		400	700	1244	47.5	1186.5	36.5
		800	1150	1248	9.9	928.6	26.7
		1600	1760	1233	12.4	843.5	16.1
32	Natural Gas	200	300	694	19.6	507.5	26.5
		500	1000	1044	28.7	949.6	32.1

3.4 Heat Flux

Schmidt-Boelter type thermopile heat flux gauges were used to continuously measure heat fluxes at five locations in the room, and one location outside the doorway of the room, cf. Table 2.5. In the interest of brevity only the heat flux measurements from a single 1/4 doorway heptanes fuel test, ISOHept9, will be analyzed. It is however important to note that these heat flux measurements only represent test ISOHept9 and that significantly different qualitative behavior in heat fluxes was observed for different fuels, different fuel distributions, and different ventilation conditions. An example of how the different fuel types affected the heat flux measurements is discussed later in section 6.1.1.

Figure 3.22 and Figure 3.23 present the heat flux measurements made in the ceiling and floor, respectively, of the enclosure for test ISOHept9. For this test the maximum peak heat flux recorded was ~300 kW/m^2. Nominally the heat flux measurements inside the room were in the range of 50 – 150 kW/m^2. As an example of how different fuels behave, the toluene fueled test ISOToluene20 had a maximum sustained heat flux of 250 kW/m^2. Initially the center ceiling heat flux measurement had a larger magnitude than the front ceiling heat flux measurement. However, halfway through the test the front ceiling heat flux measurement surpassed the center ceiling heat flux measurement. Similarly, on the floor, the rear floor heat flux measurement is initially much higher than the front floor heat flux measurement and then midway through the experiment the front floor heat flux measurement magnitude surpassed the rear floor heat flux measurement. Since both phenomena happened at approximately the same time it is reasonable to assume that the two instances are related. Considering the temperature plots presented in Figure 3.18 and Figure 3.19 it is evident that there is a development in the thermal environment of the room with the temperatures and heat fluxes in the rear half of the room tending to decrease and the temperature and heat fluxes in the front of the room tending to increase. These different trends manifest themselves immediately once the room becomes under-ventilated, cf. Figure 3.27. The likely cause for the variation in thermal behavior is therefore the under-ventilated environment of the room. Prior to becoming under-ventilated the fire has sufficient air to burn and burns more in the rear of the room as the incoming air flows over the fuel source and brings the fire to the back. As the room becomes under-ventilated there is insufficient air in the room, the fire in the rear of the room becomes very rich, and its temperature is decreased as it is diluted by excess fuel. By contrast the front of the room then contains more unburned and partially burned hydrocarbons which begin to burn in the air just as the air is entering the room. Therefore, as this situation develops the rear of the room is producing less heat at a lower temperature and the front of the room produces more heat and consequently a higher temperature.

Additionally, high heat fluxes can result in a faster vaporization of fuel. This may be evident in the heat flux measurements taken from ISOPropD14. In this test the fuel is located in two burners, one in the center of the room and the second in the rear of the room, cf. Figure 2.5. In this test the load cells attached to the burner pans revealed that the rear burner, in position 2, lost its fuel faster than the center burner, at position1. Additionally, as shown in Figure 3.24, the heat being transferred to the rear burner is so much more than the center burner that it is visibly hotter, glowing red immediately after the test, while the center burner is not. This observation is supported by the heat flux measurements presented in Figure 3.25 and Figure 3.26. The rear floor is receiving by far the greatest amount of heat flux up until the point where the fire starts to burn out. At this point the fuel in the rear burner is exhausted so burning in the front of the room is likely to produce more heat flux, however the heat flux in the rear of the room still remains quite high.

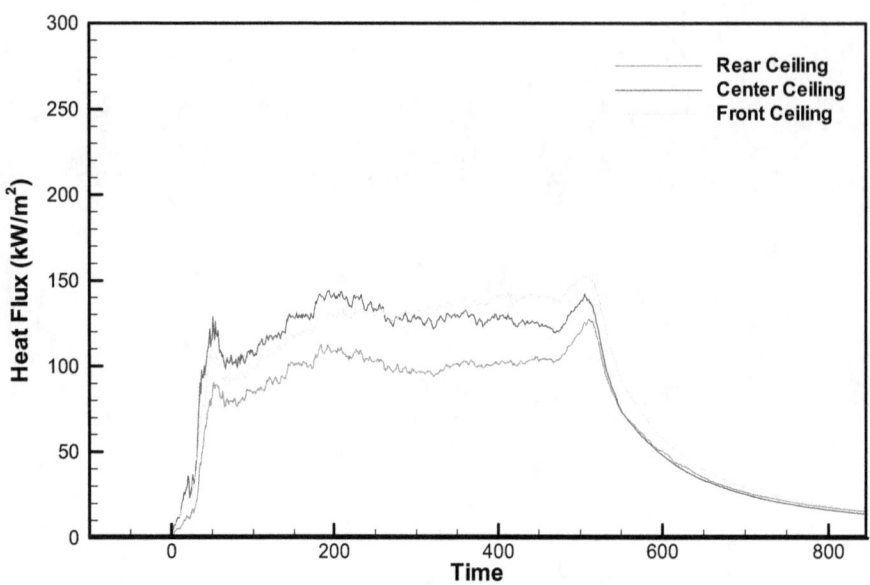

Figure 3.22: Comparison of heat flux measurements made in the ceiling for 1/4 width (20 cm) doorway heptanes fuel test ISOHept9.

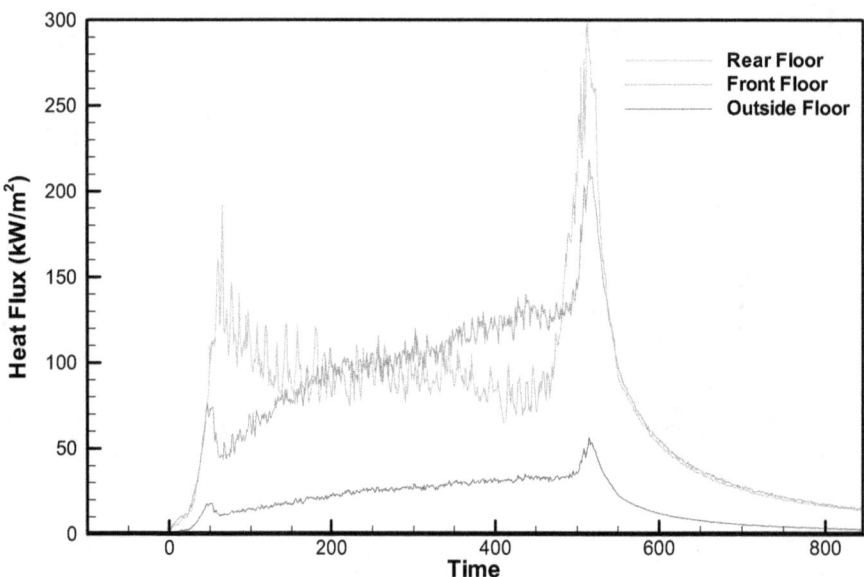

Figure 3.23: Comparison of the heat flux gauges positioned in the floor for 1/4 doorway (20 cm) heptanes fuel case ISOHept9.

Figure 3.24: Photograph of the center (front) and rear burners immediately after fire test ISOPropD14.

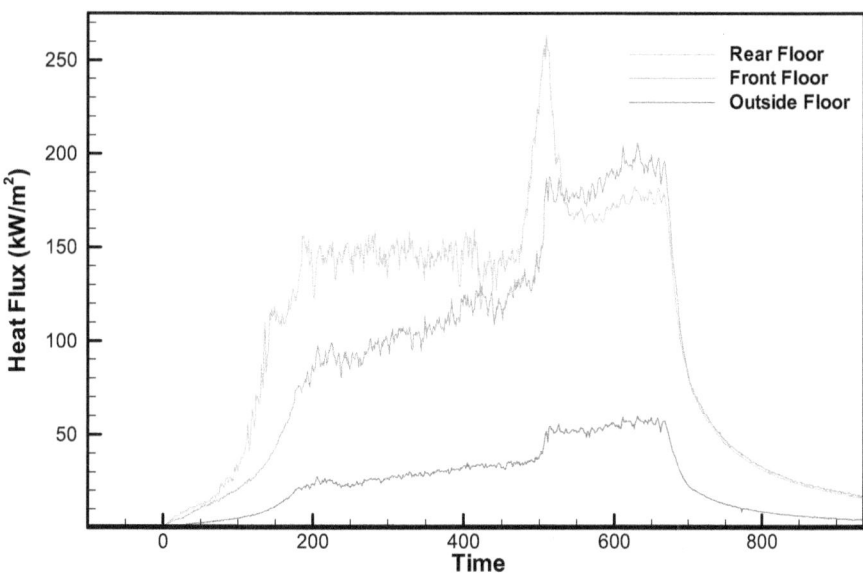

Figure 3.25: Comparison of heat flux gauge measurements for heat flux gauges positioned in the floor for 1/4 doorway (20 cm) distributed fuel isopropanol case ISOProp14.

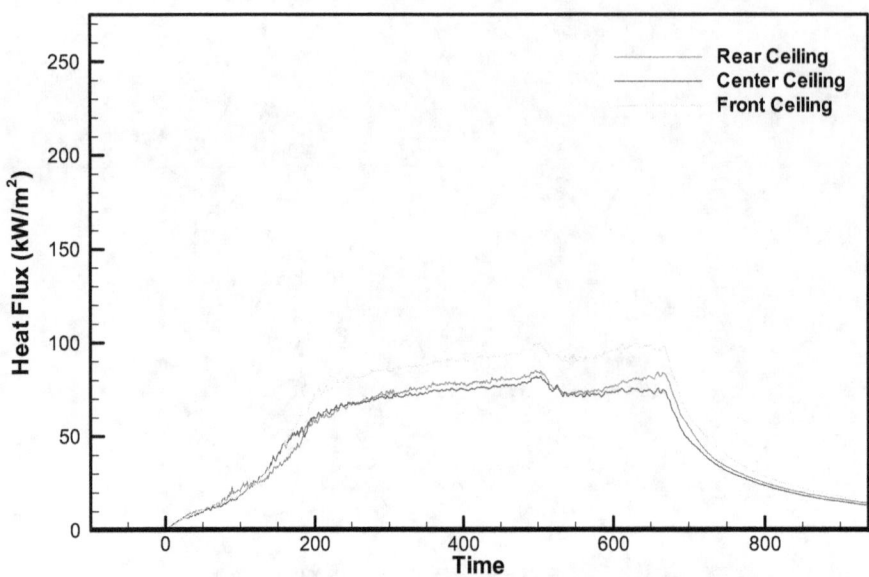

Figure 3.26: Comparison of heat flux gauge measurements for heat flux gauges positioned in the floor for 1/4 doorway (20 cm) distributed fuel isopropanol case ISOProp14.

3.5 Interior Compartment Gas Species

All gas species measurements are reported on a dry basis unless otherwise stated. Gas species O_2, CO, CO_2, and total hydrocarbons (THC) were all monitored continuously by the gas analyzers discussed in section 2.2.2. Additionally, gas chromatography (GC) measurements of H_2, O_2, CO, CO_2, N_2, CH_4, C_3H_8, heptanes, hexanes, and other higher hydrocarbons were made as well. The GC measurements were generally made with a frequency of every 2 minutes at the front sample location, and a variable frequency at the rear location due to the limited capacity of the sample storage system.

Figure 3.27 presents the transient gas species profile for the rear sample location of ISOHept9. The lines indicate gas analyzer data and the symbols indicate GC data. The agreement between the GC and gas analyzers is very good for O_2, CO, and CO_2, the three species the two systems both detect. Methane is slightly lower than the value reported by the THC analyzer, however the THC may be measuring species besides methane to report its volume fraction. The GC data also provides an additional piece of valuable information in that the volume of H_2 in the room is increasing substantially during the test indicating there is a significant amount of potential fuel in the room that was not otherwise accounted for. When H_2 is not accounted for, it also contributes to errors in the water correction.

Figure 3.28 presents measurements of gas analyzers and GC for an under-ventilated polypropylene fire, ISOPP18. Similarly to test ISOHept9, there is good agreement between gas analyzer and GC measurements for O_2, CO, and CO_2. Coincidentally the volume fractions of THC and CH_4 match for this test as well; however, they are both zero which was common to many of the solid fuels tested here.

Figure 3.29 presents the gas analyzer and GC measurements for test ISOHept27, burning heptanes fuel with a 20 cm doorway. This test is discussed in detail later in section 6.4 and is used here to illustrate some of the behaviors of the gas analyzers and GC. Considering CO, CO_2, and O_2 the gas analyzers and GC again show good agreement throughout the experiment. However, as the heat release rate increased there was an increase of total hydrocarbons in the room. Qualitatively a similar trend was observed in the volume fraction of CH_4 as it increased with heat release rate. However, the disparity between the THC and CH_4 volume fractions increased heat release rate. Therefore, early in the burn when the heat release rate was low most of the hydrocarbon species in the room were simple hydrocarbons such as CH_4. Then as the room becomes more under-ventilated larger hydrocarbons were produced. Also, large quantities of H_2 were produced in the fire, which if not accounted for can contribute to erroneous interpretation and calculations of derived quantities such as mixture fraction and water. Other than CH_4 the GC detects other higher hydrocarbons as well. Figure 3.30 presents example concentrations taken from test ISOHept27, illustrated in Figure 3.29, at a time step of 1375 seconds. Here the large concentration of H_2 is evident as well as CH_4. Additionally, small amounts of ethylene, C_2H_4, and acetylene, C_2H_2, are present which can contribute to the larger detected total hydrocarbon volume fraction reported by the analyzer. Overall it is important to note the trends in the gas species volume fractions as a function of heat release rate.

Figure 3.31 and Figure 3.32 present the steady state averages of oxygen volume fraction from the front and rear sample positions, respectively, plotted as a function of heat release rate. As the heat release rate increased there was a linear relationship with the volume fraction of O_2.

Measurements from several different fuels, including heptanes, polystyrene, toluene, polypropylene, and natural gas all exhibited nearly identical trends. This figure includes all of the experiments done in this series and nearly all of the experiments have an identical ventilation condition, 1/4 (20 cm) doorway. The exception to this is the heptanes fires where various doorway widths were compared in order to evaluate their effect, as discussed later in section 6.3. This accounts for the small amount of scatter present in the heptanes data.

Figure 3.33 and Figure 3.34 present the steady state CO_2 volume fraction measurements as a function of heat release rate for the front and rear sample locations, respectively. Initially, for heat release rates generally less than 1000 kW, the same linear trend observed in Figure 3.31 and Figure 3.32 for oxygen was observed. The variation from this linear behavior is again primarily represented by heptanes and is a result of the various ventilation conditions that were tested with heptanes. However, while the O_2 volume fraction drops linearly to zero and stays there, the CO_2 volume fraction appears to exhibit a maximum in almost all of the fuels in nearly the same place. This is similar to the maximum observed in Figure 3.29 for ISOHept27, the heat release rate ramp and correlates with the point at which most of the O_2 volume fractions in Figure 3.31 and Figure 3.32 reach zero. At this point the combustion is becoming less efficient as there is insufficient O_2 to provide complete combustion products and it is likely that other intermediate combustion products become more prevalent.

Figure 3.35 and Figure 3.36 present the steady state CO volume fraction measurements as a function of heat release rate for the front and rear sample locations, respectively. Again, similar to the CO_2 and O_2 measurements, nearly all of the measurements report the same trend as a function of heat release rate. The production of CO in the fires tends to begin to pick up at approximately 1000 kW, the same place that the O_2 volume fraction drops off and the CO_2 volume fraction reaches a maximum. For all of these plots, once all of the O_2 is consumed there is a larger variation in the results so it is not readily evident from these plots if the CO volume fraction reaches a maximum or will continue to rise. Referring back to Figure 3.29 again, the heat release rate ramp test tends to indicate that there may be a maximum in CO volume fraction after which the CO volume fraction would become lower again. This concept is expanded upon later in section 6.4.

Figure 3.37 and Figure 3.38 present the steady state total hydrocarbon (THC) volume fraction measurements as a function of heat release rate for the front and rear sample locations, respectively. The only trend evident from these plots is that once total hydrocarbons begin to be detected they continue to be produced in larger quantities as the heat release rate is increased. Generally, the THC volume fraction begins to be measureable at the same time that oxygen is being consumed and CO starts to be produced. This is the case for the liquid and gaseous fuels, however the solid fuels do not appear to produce measureable hydrocarbons at these sample locations.

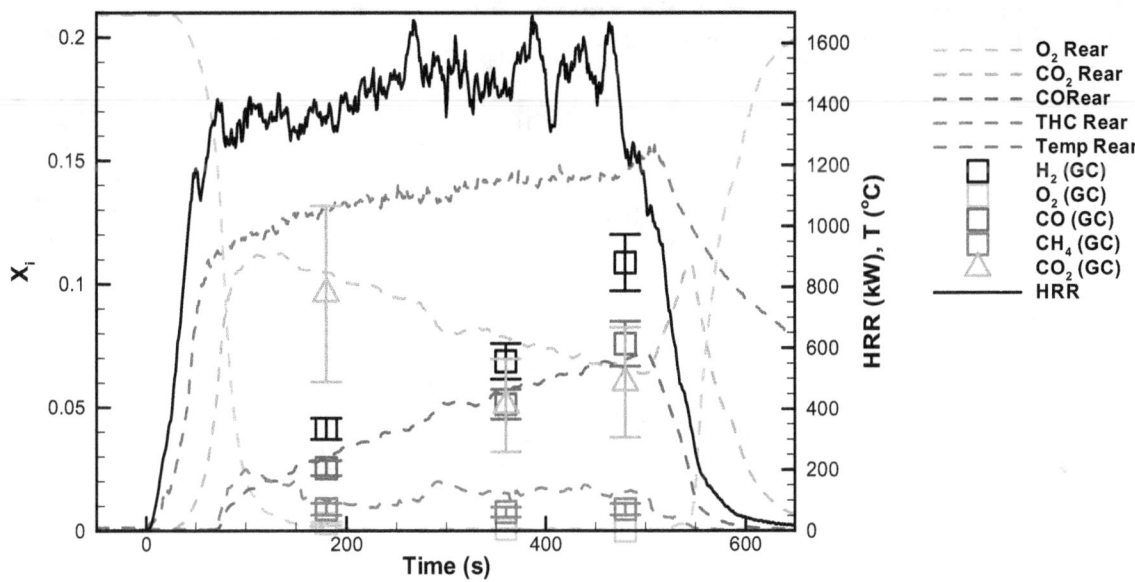

Figure 3.27: Transient gas volume fractions and soot mass fraction of test ISOHept9 (Heptane).

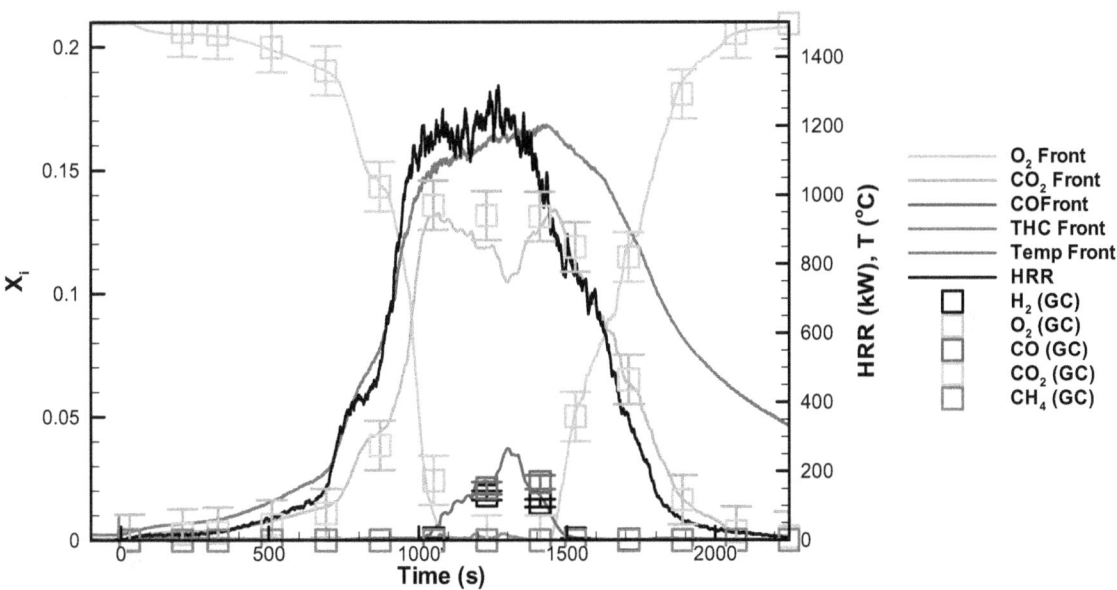

Figure 3.28: Transient gas volume fractions and soot mass fraction of test ISOPP18 (Polypropylene).

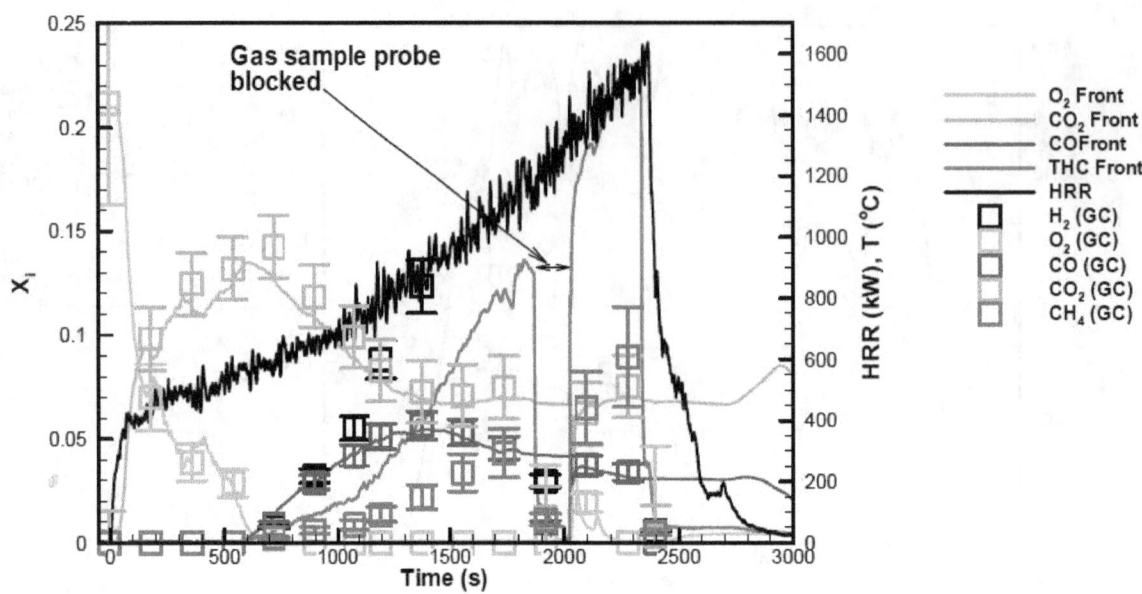

Figure 3.29: Transient gas volume fractions and heat release rate of test ISOPP18 (Polypropylene).

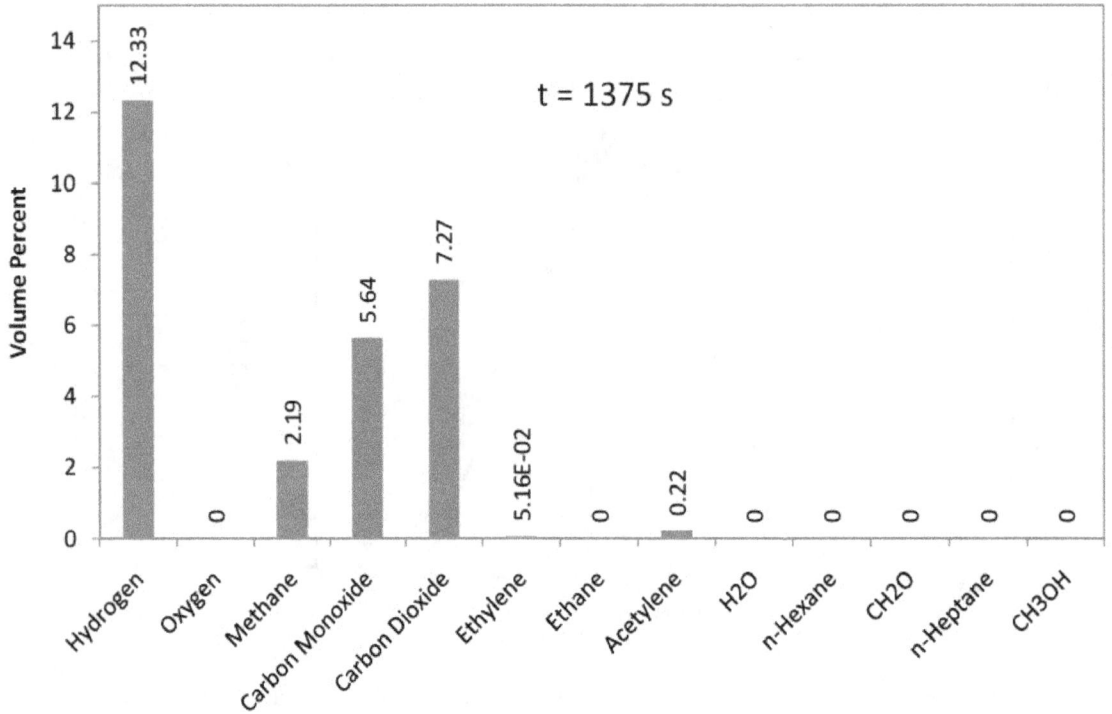

Figure 3.30: Gas species volume fractions from GC analysis of front sample location in ISOHept27 at t=1375 s.

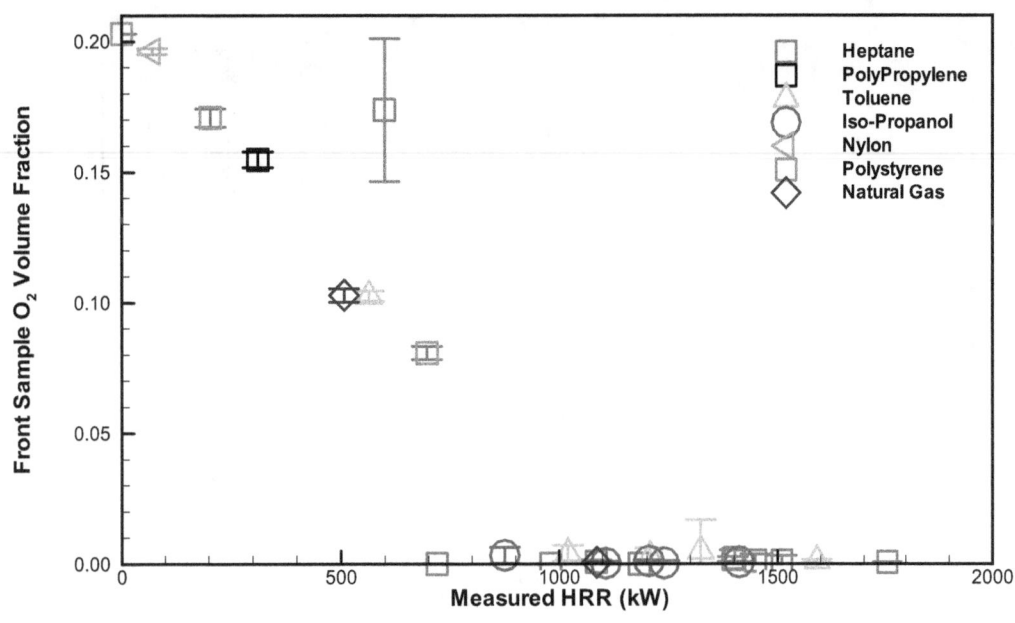

Figure 3.31: Steady state average oxygen volume fraction measurements at front sample probe location.

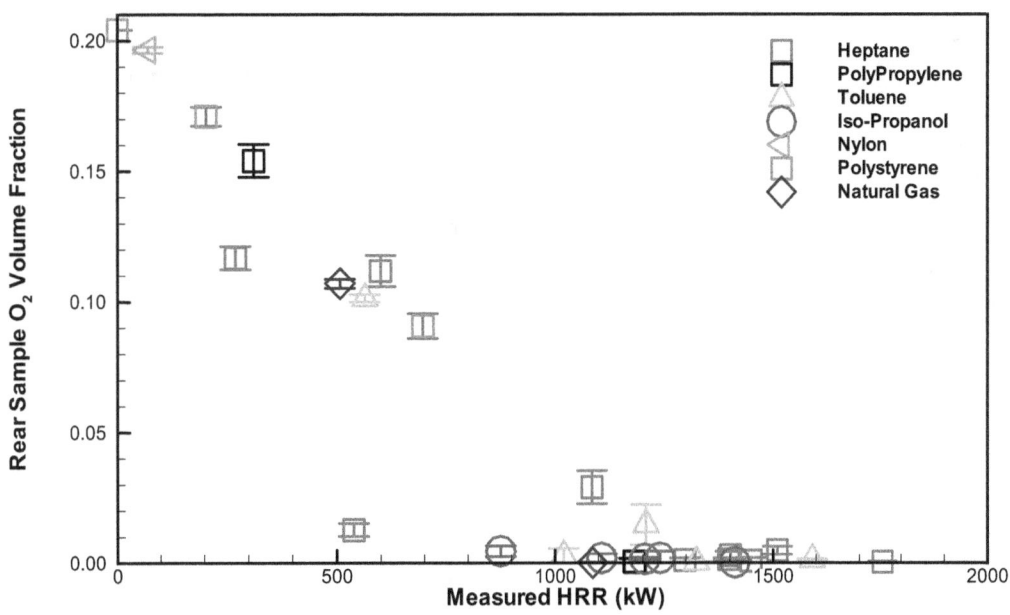

Figure 3.32: Steady state average oxygen volume fraction measurements at rear sample probe location.

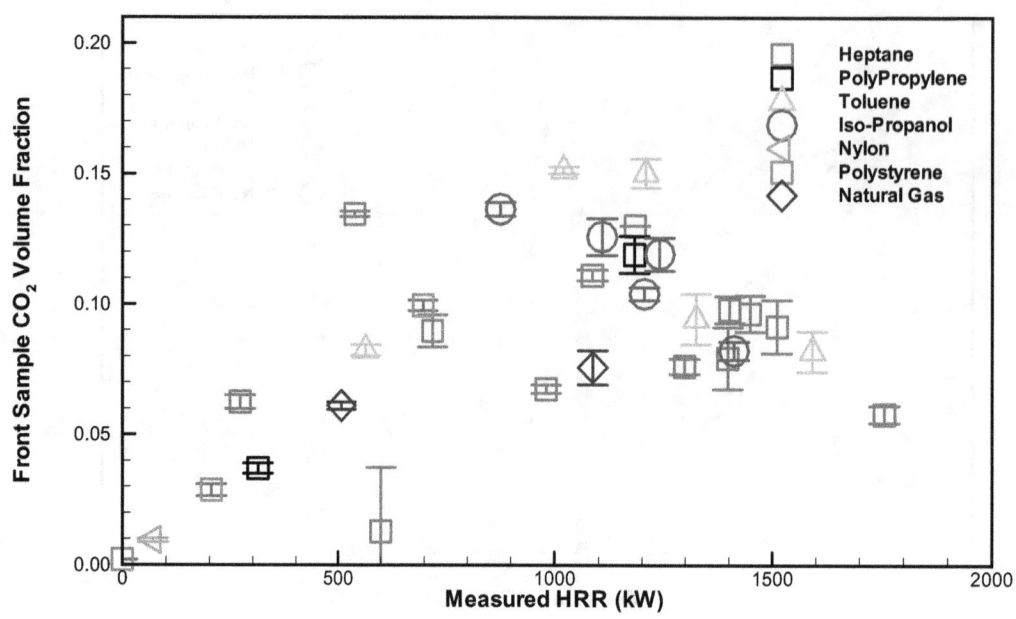

Figure 3.33: Steady state average CO_2 volume fraction measurements at front sample probe location.

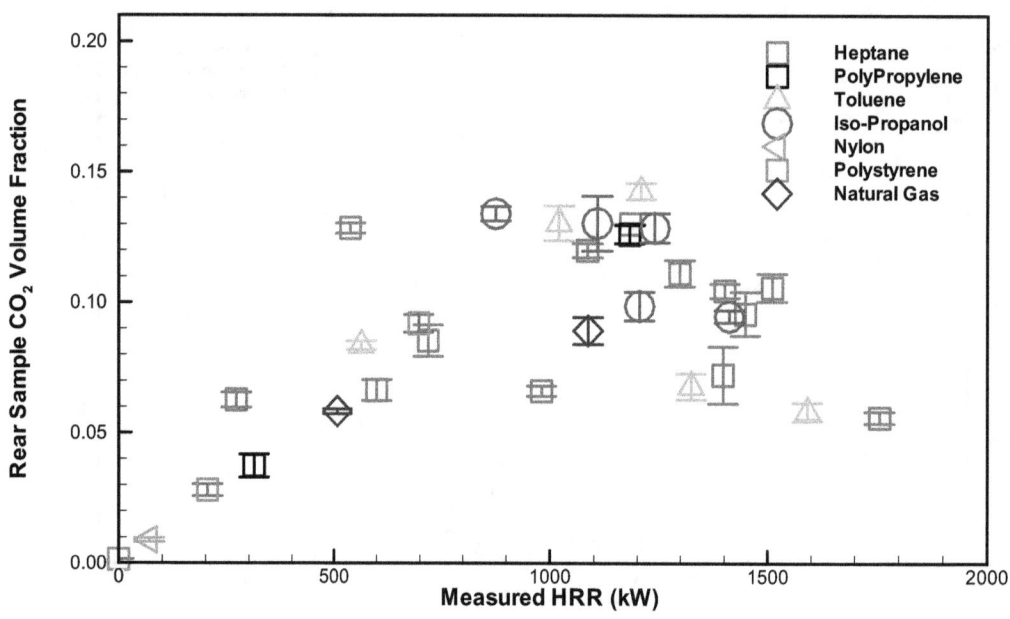

Figure 3.34: Steady state average CO_2 volume fraction measurements at rear sample probe location.

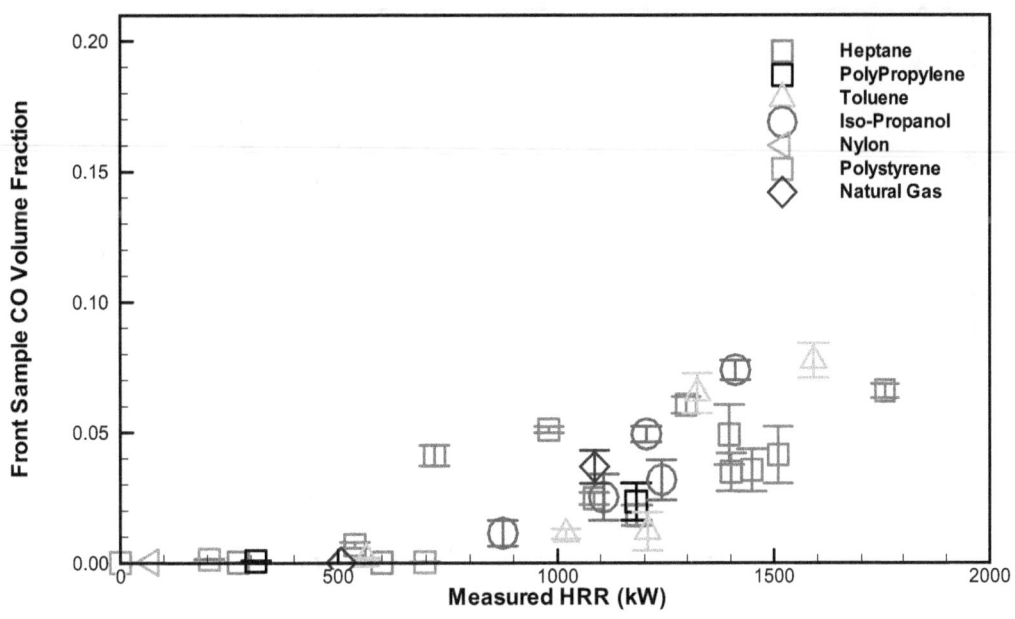

Figure 3.35: Steady state average CO volume fraction measurements at front sample probe location.

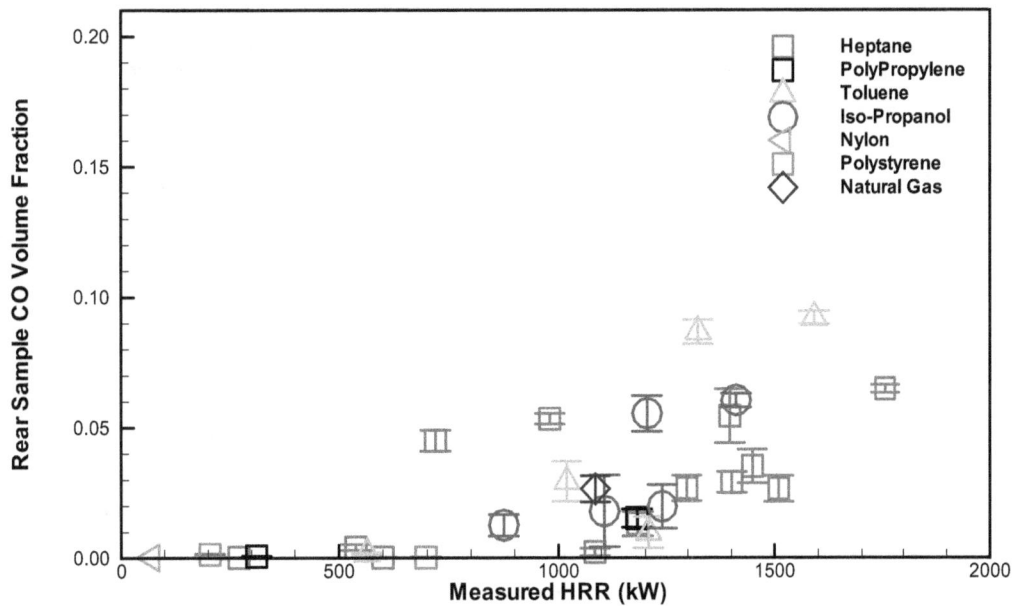

Figure 3.36: Steady state average CO volume fraction measurements at rear sample probe location.

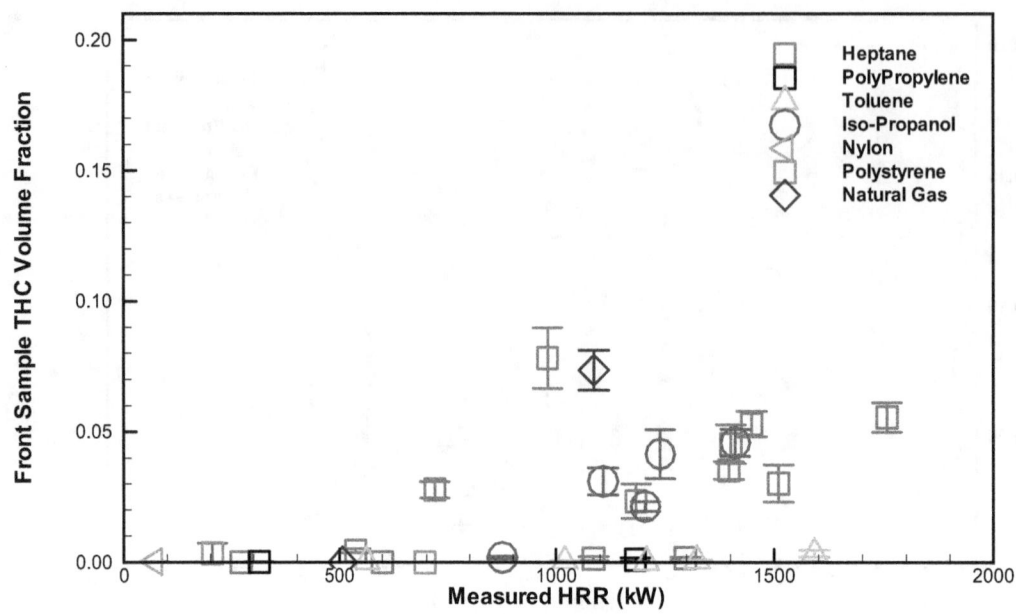

Figure 3.37: Steady state average THC volume fraction measurements at front sample probe location.

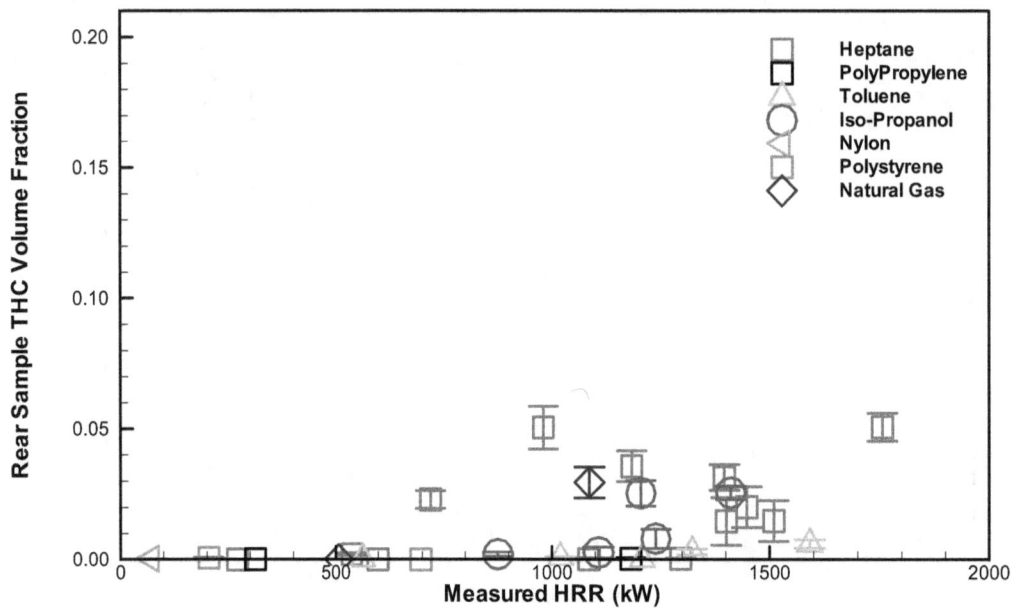

Figure 3.38: Steady state average THC volume fraction measurements at rear sample probe location.

3.6 Soot

3.6.1 Gravimetric

Soot samples were collected during 1 min to 5 min sample time after the heat release rate was quasi-steady. Measurements at the front location were conducted by the gravimetric soot probe seen in Sec. 2.2.4.1 at the front locations, while soot samples at the rear location were collected from the filter mounted at the real time extractive soot probe seen in Sec. 0.

Figure 3.39 and Figure 3.40 present the steady state gravimetric soot mass fraction measurements as a function of heat release rate for the front and rear sample locations, respectively. Overall trend shows that the soot mass fraction increases with the heat release rate, except for heptane fires that larger values are shown from 500 to 1000 kW than those over 1000 kW. This is because the heptane fire experiments were conducted under various conditions, such as burner types and door configurations (see Table 3.1). The maximum soot fractions reach to 7 % for the heptane and the toluene fires, while those do not exceed 2 % for natural gas fires. More detail on the species mass fraction results are examined further in Sec. 4.1 of this report.

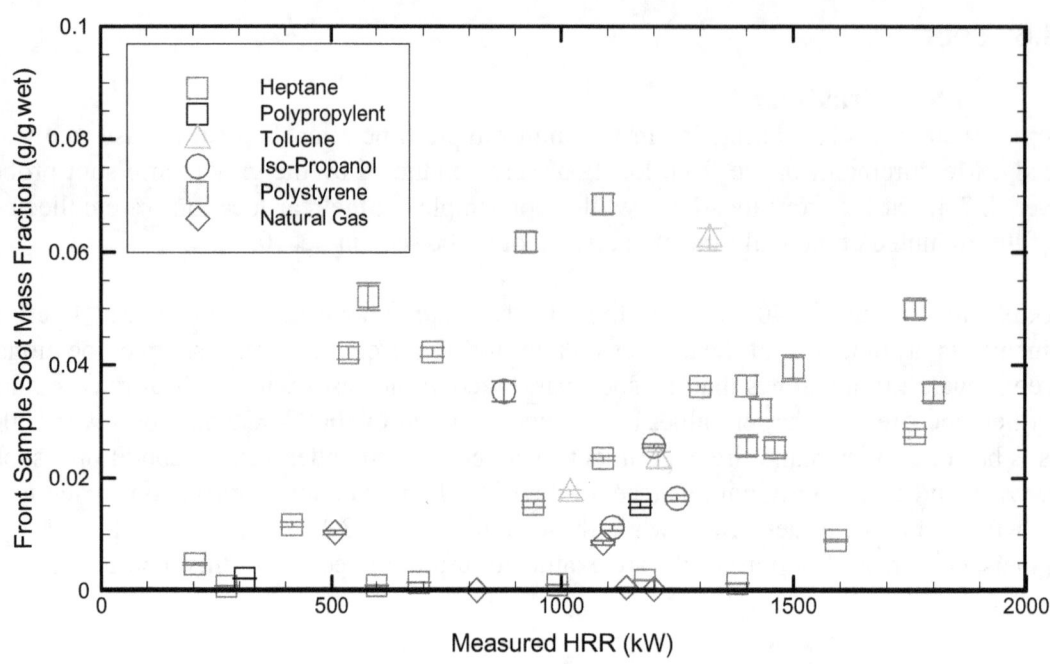

Figure 3.39: Steady state gravimetric soot mass fraction measurements at front sample probe location.

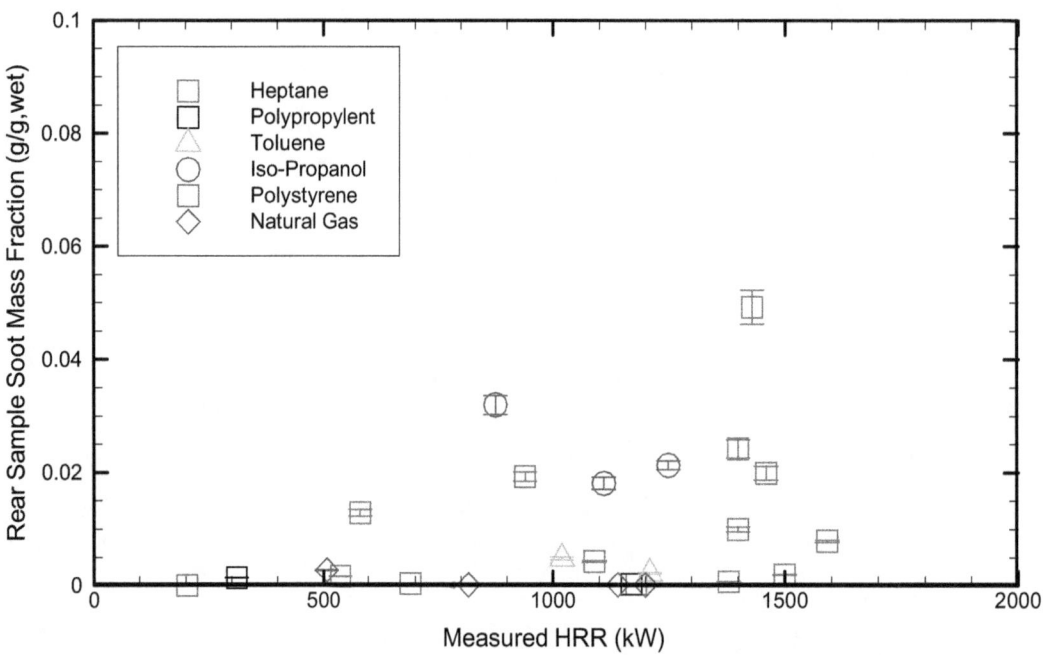

Figure 3.40: Steady state gravimetric soot mass fraction measurements at rear sample probe location.

3.6.2 Real time extractive

Results are available for Tests 21, 26, and 28 only, mainly because the measurement proved to be very challenging. Changes were made to the experimental setup throughout the test series, however due to the difficult nature of conditions inherent in this measurement, meaningful data is available only for a limited set of tests. Therefore, this data is presented as illustrative of a promising new technique that still needs further refinement.

Figure 3.41 shows a chart of the optically measured soot mass concentration plotted as a function of time for test #21, for which the fuel was polystyrene. The chart also shows the heat release rate, which had a peak value below 800 kW. The increase in soot concentration lagged behind the heat release rate; the rise in soot concentration began as the HRR approached 300 kW. Overall, the soot concentration does not appear to correlate well with the HRR. Figure 3.42 shows a comparison between the optical and gravimetric soot mass fraction in the rear of the compartment during test #21. The average optical measurement during the three minute gravimetric sampling period was 8.8×10^{-4} g/g. This was nearly seven times higher than the gravimetric average of 7.2×10^{-5} g/g.

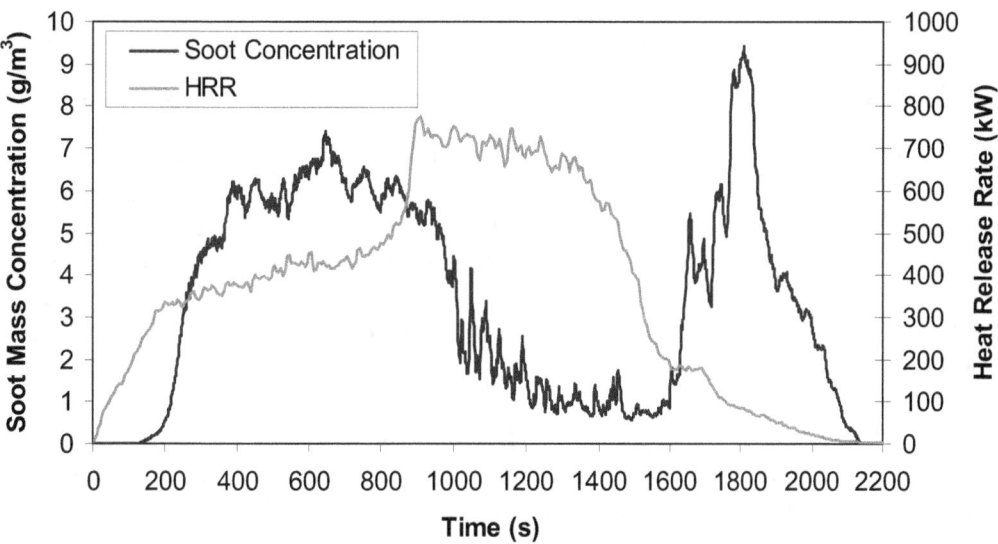

Figure 3.41: Soot mass concentration and heat release rate during polystyrene fire test 21.

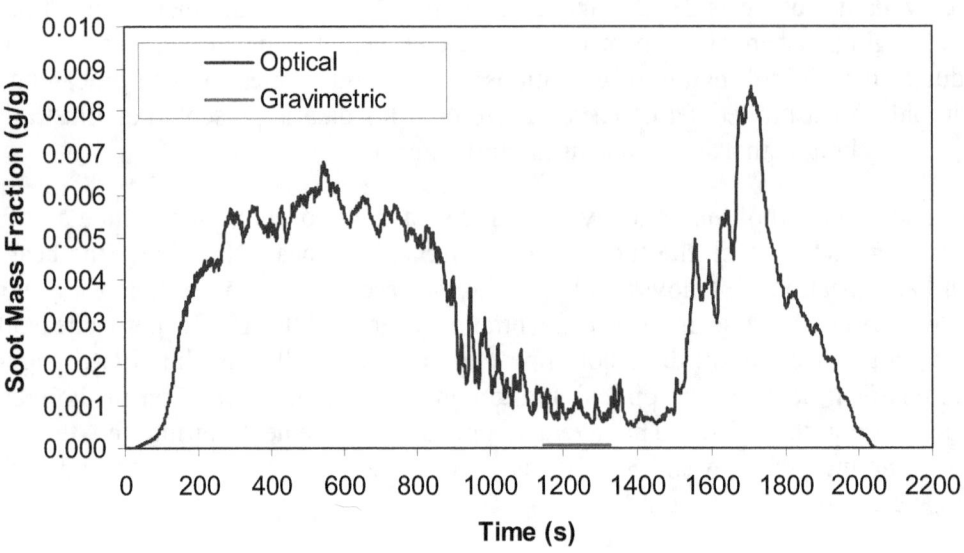

Figure 3.42: Comparison of the optical and gravimetric soot mass fraction at the rear of the compartment in test #21.

Figure 3.43 and Figure 3.44 show charts of the optically measured soot mass concentration and heat release rate for test #26 and test #28, respectively. The fuel in each of these tests was heptane. The soot mass concentration appears to track better with the heat release rate during these tests that in the polystyrene fueled test. In both tests the heat release rate was ramped initially to approximately 500 kW. At this level, the soot concentration rose steadily, but remained low. Following this period, the fire was ramped up - to approximately 1800 kW in test #26 and 1400 kW in test #28.

In test #26 the lag between the increase in HRR and the increase in soot concentration was approximately 60 seconds. After the HRR was leveled off, the soot concentration fluctuated around an average value of approximately 25 g/m^3. After the HRR was ramped down to 500 kW, the drop in soot concentration lagged by 30 seconds. After this time, the soot concentration returned to the level it had stabilized during the initial 500 kW fire.

In test #28, when the fire size was increased from 500 kW to 1400 kW, the increase in soot concentration lagged by 30 seconds. In this test, with the HRR stabilized, the soot concentration was not as consistent. It climbed to 27.5 g/m^3 at 300 seconds, then dropped to 8.3 g/m^3 at 580 seconds. This was followed by a rapid increase to 48.7 g/m^3 at 590 seconds. The concentration then dropped off for the remainder of the test, despite relatively stable heat release rate levels.

Figure 3.45 shows a comparison between the optical and gravimetric soot mass fraction in the rear of the compartment during test #28. The average optical measurement during the two minute gravimetric sampling period was 0.0138 g/g. This was nearly seven times higher than the gravimetric average of 0.002 g/g.

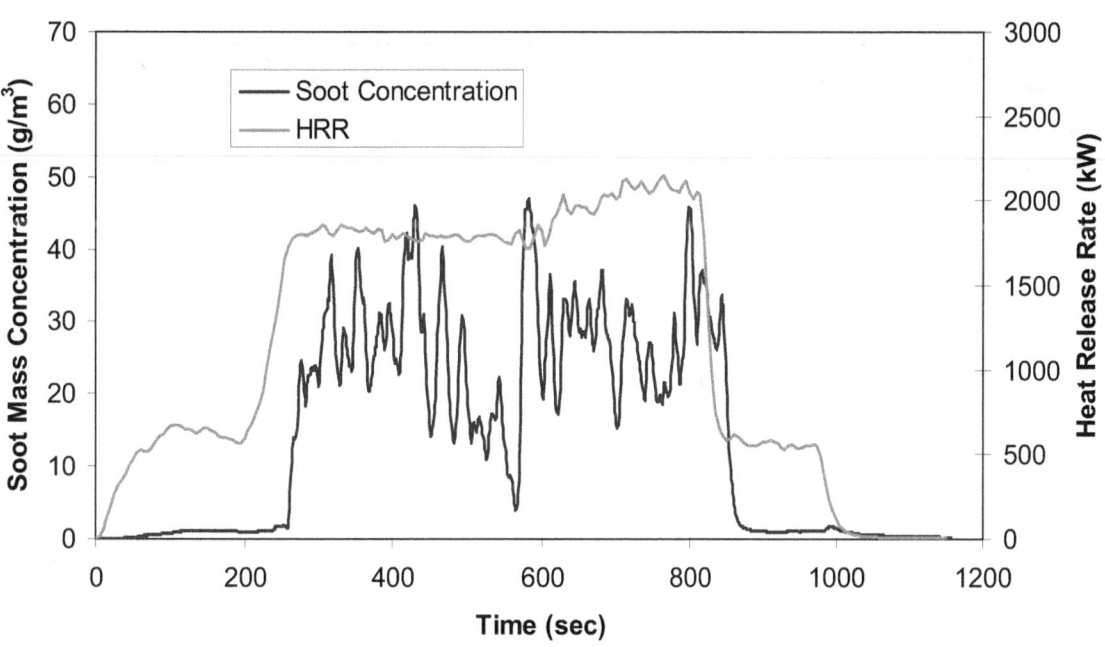

Figure 3.43: Soot mass concentration and heat release rate during heptane fire test 26.

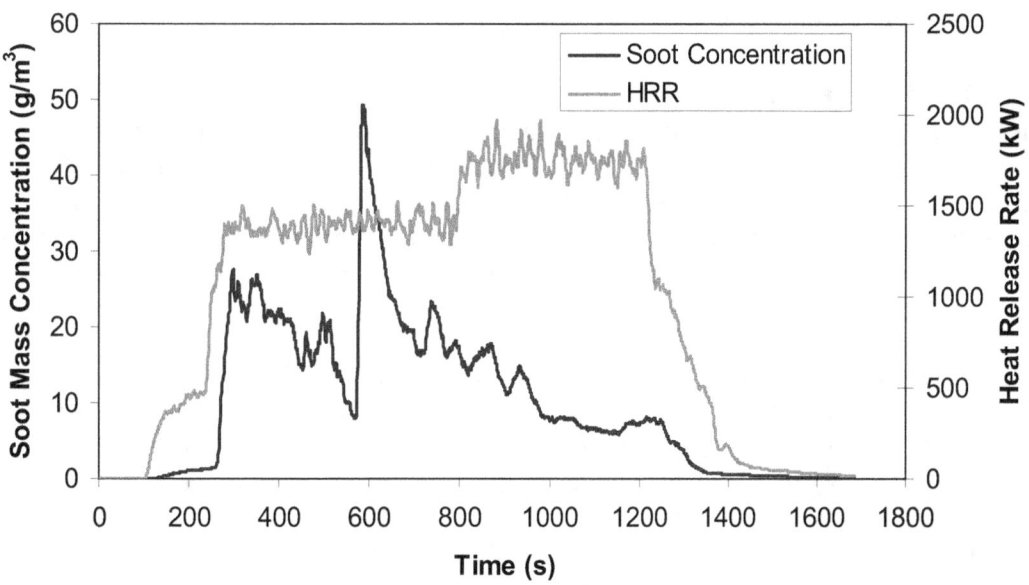

Figure 3.44: Soot mass concentration and heat release rate during heptane fire test 28.

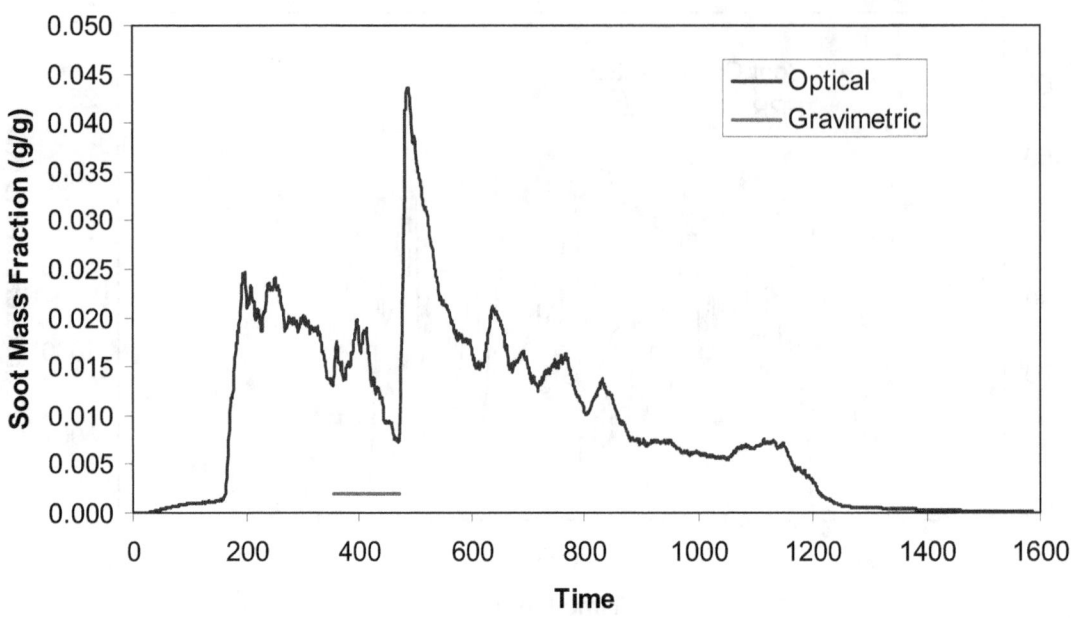

Figure 3.45: Comparison of the optical and gravimetric soot mass fraction at the rear of the compartment in test #28.

4 COMPARTMENT CHEMISTRY ANALYSIS

4.1 Mixture Fraction Analysis

It is useful to consider the compartment fire composition measurements in terms of the mixture fraction. The use of mixture fraction to analyze flame data was first used by Bilger [39] and later modified by Peters [40] and others. The mixture fraction approach has been widely used to represent the chemistry in turbulent flame models and fire field models, and has been used to analyze the structure of laminar counterflowing and coflowing hydrocarbon and alcohol flames [41, 42]

Pool fires and compartment fires differ from simple laminar flames, as they are typically transient and turbulent by nature. Yet, application of the mixture fraction concept to these complex combustion situations can provide additional insight into the structure of the fire. The mixture fraction approach allows evaluation of a set of species measurements in terms of self-consistency, and at the same time facilitates rapid assessment of the overall behavior of a combustion system. Floyd et al [43] applied the mixture fraction approach to evaluate the species composition at various locations in compartment fires. Pitts [44] measured the local equivalence ratio at various locations in compartment fires, investigating the possibility of a correlation for CO. Since there is a one-to-one correspondence between mixture fraction and equivalence ratio, the approach used here is similar to that used previously by Pitts [44] and other experimentalists, with the difference that soot is considered in the analysis of mixture fraction and local equivalence ratio.

Sivathanu and Faeth [45] considered the relationship between soot and mixture fraction in an effort to improve the understanding associated with radiative emissions from fires. Their measurements [45] clearly showed that soot did not correlate well with mixture fraction in laminar hydrocarbon diffusion flames. Their data suggest, however, a relationship between soot volume fraction and temperature in the fuel rich regions of turbulent hydrocarbon diffusion flames.

Recently, a mixture fraction analysis was performed to investigate the characteristics of chemical species production in the upper layer of the 2/5 scale compartment based on the ISO-9705 room [5]. The analysis showed that plotting the local composition as a function of the mixture fraction collapsed hundreds of species measurements from an assortment of compartment conditions, with varying heat release rates, burner types and spatial locations, into a few coherent lines or bands. Also, inclusion of soot into mixture fraction analysis allowed identification of fuel rich or under-ventilated conditions for the compartment fires of smoky fuels, such as heptane, toluene, and polystyrene. The analysis performed here for full scale experimental data is the extension of our previous study [5] for the reduced scale compartment fires.

In this section, the mixture fraction was used to evaluate the species composition in the hot upper layer of the compartment fires. The analysis provides a check on the quality of the data and provides insight into the chemistry of compartment fires. Also, the significance of the inclusion of soot as part of the mixture fraction analysis was investigated. The importance of measurement uncertainty is highlighted, and its value is quantified as part of the mixture fraction analysis.

4.1.1 Definition of Mixture Fraction

The mixture fraction is a non-dimensional quantity representing the mass fraction of a species, at a particular location, that was originally part of the fuel stream. The mixture fraction based on carbon containing species is defined as follows:

$$Z = Y_F + Y_{co} \frac{MW_F}{x\,MW_{co}} + Y_{co_2} \frac{MW_F}{x\,MW_{co_2}} + Y_{Soot} \frac{MW_F}{x\,MW_{Soot}} \qquad 13$$

or

$$Z = Y_{UH} \frac{MW_F}{x\,MW_{UH}} + Y_{co} \frac{MW_F}{x\,MW_{co}} + Y_{co_2} \frac{MW_F}{x\,MW_{co_2}} + Y_{Soot} \frac{MW_F}{x\,MW_{Soot}} \qquad 14$$

where MW_i is the molecular weight of chemical species i, Y_i is the mass fraction of that species, x is the number of carbon atoms in the parent fuel molecule ($C_x H_y O_z$), MW_F is the molecular weight of the parent fuel, MW_{CO} is 28 g/mol, MW_{Soot} is taken as 12 g/mol (assuming that soot can be approximated as pure carbon), and MW_{CO_2} is 44 g/mol. Alternative definitions of mixture fraction yield results similar to those shown below.

In the experiments reported here, the measurement of unburned hydrocarbon (UH) was made using the total hydrocarbon analyzer, reported on an equivalent methane (CH$_4$) basis. Thus, Eq. 14 is used to calculate mixture fraction from experimental data instead of Eq.13.

In the fire literature, soot is typically not considered in Eq. 13. Here, it is included formally. In the analysis given below, the results with and without soot in Eq. 13 and 14 are compared with each other. Its inclusion is especially important for highly sooting conditions, as will be shown in the results section below.

The mass fraction, Y_i, of each species i is determined from the measured volume fraction, X_i, by the following expression:

$$Y_i = X_i MW_i / MW_{tot} \qquad 15$$

MW_{tot} represents the average molecular weight of all gas species and is a function of the local composition.

$$MW_{tot} = \sum_i X_i MW_i \qquad 16$$

The state relations can be derived by considering the idealized reaction of a hydrocarbon fuel, rewritten here in an expanded form of Eq. 7 (in Sec. 2.5):

$$\begin{aligned}C_xH_yO_z + \eta(x+y/4-z/2)(O_2 + 3.76N_2) \rightarrow \max(0, 1-\eta)C_xH_yO_z + \min(1,\eta)xCO_2 \\ + \min(1,\eta)(y/2)H_2O + \max(0, \eta-1)(x+y/4-z/2)O_2 + \eta(x+y/4)3.76N_2\end{aligned} \qquad 17$$

where the function $\max(\alpha,\beta)$ returns the larger of the two parameters, α or β, and the function $\min(\alpha,\beta)$ returns the smaller of the two parameters, α or β. Here, η is a parameter ranging from zero (all fuel and zero oxygen) to infinity (all oxygen and zero fuel) and becomes unity for stoichiometric conditions. The definition of η shows that it is the reciprocal of the local fuel equivalence ratio, ϕ.

$$\phi = \frac{(F/A)}{(F/A)_{st}} = \frac{MW_F / \eta(x+y/4-z/2)(MW_{O_2} + 3.76 MW_{N_2})}{MW_F / (x+y/4-z/2)(MW_{O_2} + 3.76 MW_{N_2})} = \frac{1}{\eta} \qquad 18$$

where F/A is the fuel-air ratio and the subscript 'st' refers to stoichiometric conditions. The idealized mass fractions of products are obtained from the right side of the Eq. 17. At the flame sheet where both the fuel and oxygen concentrations go to zero, $Y_F = Y_{CO} = Y_{soot} 0$, and Eq. 14 leads to:

$$Z_{st} = Y_{CO2} \frac{MW_F}{x \, MW_{CO2}} \qquad 19$$

The value of the stoichiometric mixture fraction for the fuels considered in this report is shown in Table 4.1. Its value varied from about 0.0554 for natural gas to 0.1346 for methanol.

Table 4.1. Stoichiometric Value of the Mixture Fraction (Zst) for different fuels.

Fuel	Chemical Formula	Z_{st}
Methane	CH4	0.0552
Natural Gas	0.93 CH4 + 0.04 C2H6 + 0.01 C3H8 + 0.01 CO2 +..[*]	0.0554 ± 0.0002[**]
n-Heptane	C7H16	0.0622
Toluene	C7H8	0.0694
Polystyrene	(C8H8) n	0.0705
Polypropylene	(C3H6)n	0.0637
Iso-Propanol	C3H7OH	0.0885

[*] typical composition; actual composition varies day to day.
[**] average value based on measured natural gas composition.

A mixture fraction calculation for a methane-air flame is presented here as an example. For methane, Eq. 17 becomes:

$$CH_4 + 2\eta(O_2 + 3.76N_2) \rightarrow \max(0, 1-\eta)CH_4 + \min(1,\eta)CO_2 + 2\min(1,\eta)H_2O \\ + 2\max(0, \eta-1)O_2 + 7.52\eta N_2 \qquad 20$$

The traditional mixture fraction model holds that the mass fraction, Y_i, of products can be determined through the right side of Eq. 20 as follows:

$$\begin{aligned}
Y_{CH_4} &= \max(0, 1-\eta) MW_{CH_4} / MW_{tot} \\
Y_{CO_2} &= \min(1, \eta) MW_{CO_2} / MW_{tot} \\
Y_{H_2O} &= 2\min(1, \eta) MW_{H_2O} / MW_{tot} \\
Y_{O_2} &= 2\max(0, \eta-1) MW_{O_2} / MW_{tot} \\
Y_{N_2} &= 7.52\eta MW_{N_2} / MW_{tot}
\end{aligned} \qquad 21$$

where Y_{CO} and Y_{Soot} in Eq. 14 were taken as zero for this mixture fraction model calculation. The molecular weight of the mixture is a function of the local composition and can be calculated from the reactant concentrations:

$$MW_{tot} = MW_{CH_4} + 2\eta(MW_{O_2} + 3.76 MW_{N_2}). \qquad 22$$

Since Y_{CO} and Y_{Soot} are assumed to be equal to zero and $Y_F = Y_{CH_4}$ the mixture fraction defined in Eq. 14 can be rewritten as:

$$Z = Y_{CH_4} + Y_{CO_2} \frac{MW_{CH_4}}{x\, MW_{CO_2}} \qquad 23$$

Using Eqs. 21 and 22, Eq. 23 can be rewritten as:

$$Z = \frac{MW_{CH_4}}{MW_{tot}} = \frac{MW_{CH_4}}{MW_{CH_4} + 2\eta(MW_{O_2} + 3.76MW_{N_2})}, \qquad 24$$

and

$$\eta = \frac{(1-Z)}{Z} \frac{MW_{CH_4}}{2(MW_{O_2} + 3.76MW_{N_2})} = \frac{(1-Z)}{Z}(F/A)_{st}. \qquad 25$$

Figure 4.1 presents the relationship between the mixture fraction and the equivalence ratio ($1/\eta$) as delineated in Eq. 5.12 for the methane-air system. Under stoichiometric conditions ($\eta = 1$), the mixture fraction is 0.0552 for a methane-air flame as listed in Table 4.1. In Figure 4.1, natural gas is treated as if it were methane. The figure shows that the mixture fraction compresses a large range of equivalence ratio values. Figure 4.2 shows the relationship between the mass fraction and the mixture fraction for most of the major species in the methane-air system, when Y_{CO} is taken as zero.

4.1.2 Mixture Fraction Uncertainty

The uncertainty in the mixture fraction is propagated through Eq. 14 and is based on the measurement uncertainty of the species concentrations. The positive square root of the estimated variance, $U_Z(Y_i)$, is obtained from

$$U_Z^2(Y_i) = \sum_{i=1}^{N} \left[\frac{\partial Z}{\partial Y_i}\right]^2 U_{Y_i}^2 \qquad 26$$

where U_{Y_i} is an estimate of the combined expanded measurement uncertainty of the measured mass fraction, Y_i, of species i.

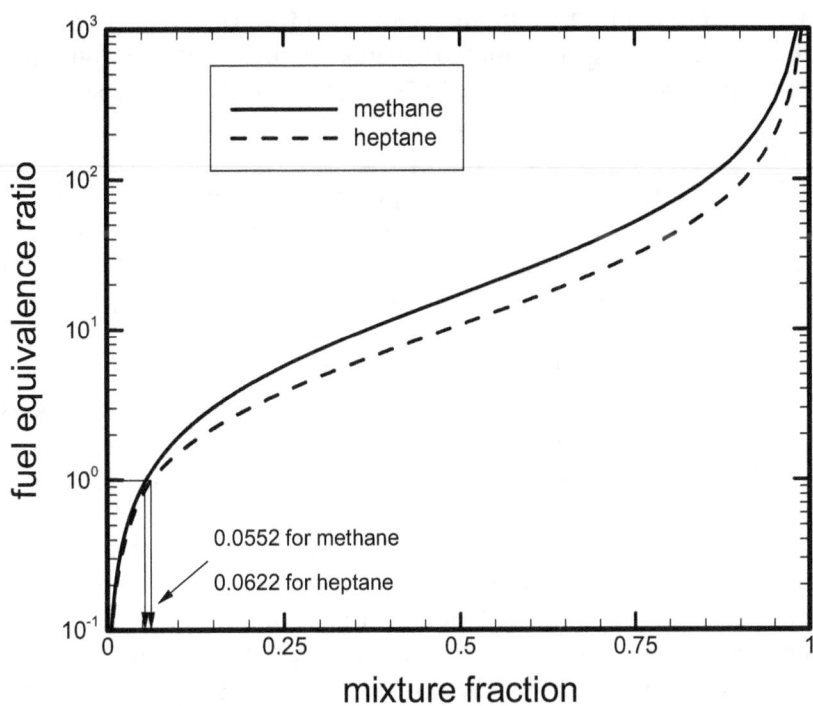

Figure 4.1: The equivalence ratio as a function of mixture fraction for nonpremixed flames burning methane and n-heptane.

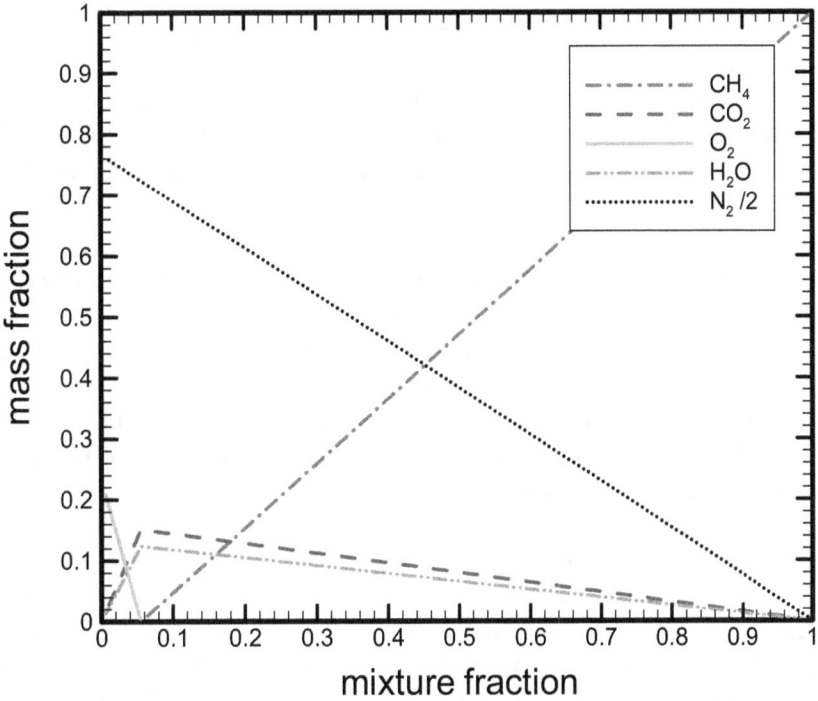

Figure 4.2: The mass fraction vs. the mixture fraction calculated by the single-parameter mixture fraction model.

4.1.3 Species Composition Results in terms of Mixture Fraction

In this section, the time-varying species measurements are presented as a function of the mixture fraction. The results are organized in terms of fuel type, since the fuel type establishes the basis for the correlation (see Eqs. 7 (in Sec.2.5) and 17).

The species data are considered in terms of the species mass fraction (Y_i), which is plotted as a function of the local mixture fraction (Z), based on the fuel mass. Measurements from the front and rear of the compartment, for all fire conditions (i.e., heat release rate, burner type) and all times during the experiment are plotted on a single graph in terms of mixture fraction. The mass fractions of H_2O and N_2 were not measured in the experiments; the values of these species in this report (and shown in Figure 4.3 through Figure 4.8) are estimated from the stoichiometric relation (Eqs. 7-12 in Sec. 2.5). The mass fractions of the unburned hydrocarbons (UH) in each plot were taken from the hydrocarbon analyzer measurements. The total hydrocarbons (UH) results were normalized in terms of the equivalent fuel molecule for each fuel type.

The lines in Figure 4.3 through Figure 4.8 represent complete stoichiometric combustion and the hypothetical case when only CO_2 is produced (no CO or soot; see Figure 4.2). In some cases, because of the number of data points, the theoretical lines are somewhat obscured. The lines on the average steady-state measurement plots labeled as "(b)" and "(c)" are easier to distinguish as the plots are less crowded. In those plots, the propagated uncertainty is also presented. Soot was not measured at all times, but only during the periods when the fire heat release rate was quasi-steady. Soot is shown only on the plots labeled "(c)" and is presented with the time-averaged gas species as a function of mixture fraction including soot.

Figure 4.3 presents all of the gas species measurements taken during the natural gas experiments (tests #2 - #3, and #32) in both the front and rear of the compartment as a function of mixture fraction. Figure 3a shows all of the transient measurements for all of the natural gas tests with the full- and quarter- door configuration (tests #2-#3, and #32, respectively). Figure 4.3b and c show the time-averaged steady-state measurements as a function of mixture fraction without and with soot, respectively. At any single location, the mixture fraction can vary from lean to rich, due to the dynamics of the fire. The stoichiometric mixture fraction (Z_{st}) is a useful reference point for consideration of fire chemistry (see Table 4.1; $Z_{st} = 0.0544$). For fuel lean conditions ($Z < Z_{st}$), the measured mass fractions of methane and carbon monoxide are near zero. As the mixture fraction increases, the mass fraction of oxygen decreases, and the carbon dioxide and water vapor mass fractions increase. For mixture fraction values greater than stoichiometric, the oxygen mass fraction approaches zero, whereas the fraction of unburned fuel increases approximately linearly. Under these conditions, the generation of carbon monoxide is observed and Y_{CO} attains a maximum value of about 0.04 g/g.

As seen in the figure, the hypothetical lines show reasonable agreement with the measurements for fuel lean and near-stoichiometric conditions. As the mixture fraction increases beyond stoichiometric, however, the difference between the hypothetical lines and the measurements becomes considerable. In Figure 4.3a, there are some data that does not follow the theoretical lines. Contrarily, the deviated points connected to these data are not shown in the time-averaged measurement plots (Figure 4.3b and c). It indicates that these data were measured when the fire has been developing and ceasing, not when the HRR was quasi-steady. The value of Y_{CO} is not negligible for fuel rich conditions. As a result, the hypothetical lines over-predict the CO_2 mass fraction by about 10 % for mixture fraction values for $Z_{st} < Z < 0.12$. As expected, the plots show that the simple traditional mixture faction approach does not correlate the experimental results for CO. This behavior is also observed in laminar flames, which is attributed to finite rate chemistry effects associated with slow CO chemistry [42]. Other approaches to predict CO, possibly using variables that are functions of mixture fraction, will need to be considered to improve predictions of its concentration.

The vertical and horizontal error bars in Figure 4.3b represent the combined expanded uncertainty of the mass fractions of gas species and the mixture fraction, respectively. The uncertainties in the mixture fraction increase with the mixture fraction, and the values of relative errors are under 10% except one point showing 22% relative error. Figure 4.3c shows the mass fractions of gas species as a function of mixture fraction calculated with soot mass fraction (Eq. 14). For fuel lean and stoichiometric conditions, the mixture fractions calculated with soot are almost same as those without soot because amounts of soot are very small. However, the gas species of Z=0.09 in Figure 4.3b move to Z=0.11 in Figure 4.3c. The soot mass fraction of this point is about 0.01 g/g. By this move of data, the deviation of unburned hydrocarbon from theoretical line increases, but the sum of unburned hydrocarbon and soot mass fraction follows the line instead.

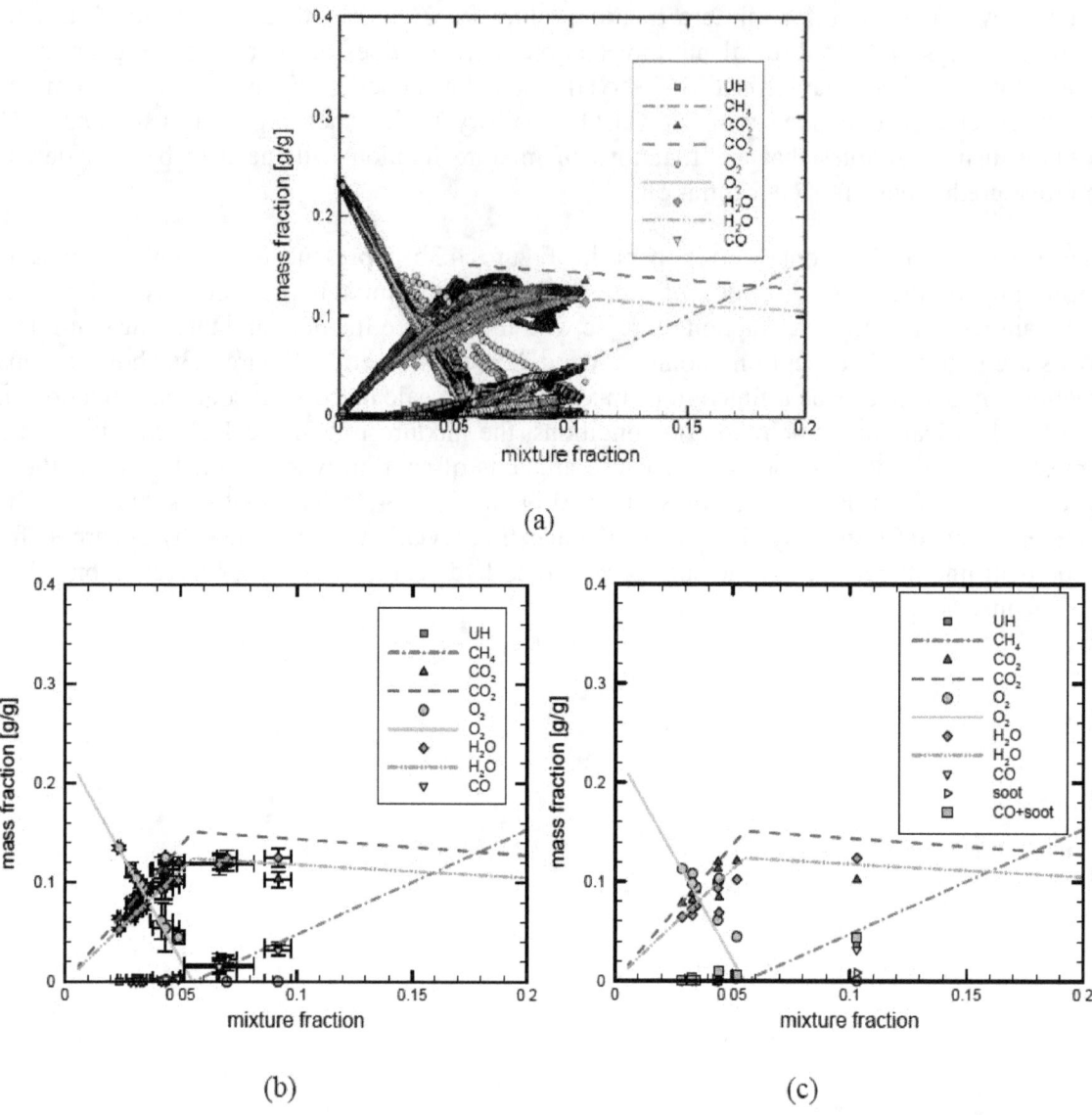

Figure 4.3: Mass fractions of front and rear compartment gas species for the natural gas fire tests #1-#3, and #32: (a) transient measurements, (b) time-averaged measurements as a function of mixture fraction without soot, and (c) time-averaged measurements as a function of mixture fraction including soot.

4.1.4 Condensed-Phase Hydrocarbon Fuels

Figure 4.4 to Figure 4.7 show the mass fraction as a function of mixture fraction for the fires burning heptane (tests #5, #9, #12, #13, #19, #22-#26, #28), toluene (tests #20, #29), polypropylene (tests #11, #18) and polystyrene (tests #16, #17, #21), respectively. These contain the species concentrations measured under various configurations, such as different burner types, numbers, sizes, and door sizes (see Table 3.1). For small values of mixture fraction ($Z << Z_{st}$), the overall trends of species mass fractions follow to the mixture fraction model for toluene, polypropylene, and polystyrene fires. Even for the heptane fires, the steady-state data agree with the theoretical lines. The deviated data in Figure 4.4a, especially for oxygen mass fraction, are measured during fire developing periods. Figure 4.4 to Figure 4.7 shows, for near-stoichiometric conditions, Y_{CO} is non-zero, which leads to Y_{CO_2} lower than predicted by the mixture fraction model. The values of Y_{CO} reach to 6 %, 9 %, 0.2 % and 0.1 % for heptane, toluene, polypropylene, and polystyrene, respectively.

Figure 4.4b, Figure 4.5b, Figure 4.6b and Figure 4.7b show that the measurement uncertainty was relatively small for lean mixture fractions. For large values of the mixture fraction, the variance of the species mass fraction results was relatively broad. This is particularly true for the transient results, but also for the time-averaged results. By Figure 4.5b, Figure 4.6b and Figure 4.7b, the local conditions are not fuel-rich for any of the conditions investigated during the toluene, polypropylene and the polystyrene tests. It is interesting because at least one case for each fuel has reached the flashover condition (see Table 3.1) and under-ventilation condition. Although there was a high concentration of mass fraction results about near-stoichiometric conditions, the negligible amounts of hydrocarbons measured during these tests led to mixture fraction values which were less than stoichiometric in value.

Species concentration results in the fire literature, such as those presented in Figure 4.4b, Figure 4.5b, Figure 4.6b and Figure 4.7b, are typically reported without consideration of soot in the definition of mixture fraction. It is correct to include soot in Eq. 23 as the conserved scalar approach is based on the idea that elemental mass is neither created nor destroyed in a fire. The appearance of the plots qualitatively changes when soot is considered. Figure 4.4c, Figure 4.5c, Figure 4.6c and Figure 4.7c show that the inclusion of soot reduces the scatter in the mass fractions for large values of Z, while otherwise leaving the plots unchanged. Inclusion of soot stretches the value of Z proportional to the measured soot mass fraction in a non-linear manner as illustrated in Figure 4.8. This is because Y_{Soot} is negligible for lean conditions, whereas it is significant for large values of Z, taking on values as large as 0.1. Neglecting soot for the fires burning natural gas may be reasonable, whereas considering it for heavily sooting fires is necessary. The scatter in the mass fractions was reduced for these fuels when soot was considered in the definition of Z (see Eq. 14). In Figure 4.4c, Figure 4.5c, Figure 4.6c and Figure 4.7c, the sum of the soot and unburned hydrocarbons (UH) appears to closely follow the mixture fraction model results. The results plotted in this way are particularly convincing in Figure 4.4c, Figure 4.5c, where the independent results for soot and THC do not follow the state relationship model, but their sum does. Interestingly, Figure 4.5b, Figure 4.6b and Figure 4.7b show that there was no significant amount of UH measured in the upper layer of the compartment in the toluene, polypropylene or polystyrene fires. The carbon in the upper layer of these fires is primarily in the form of CO, CO_2 or soot. Examination of Figure 3.23 and Figure 3.24 present the same data, which reaffirms that the total HC measurements were relatively small in the toluene, polypropylene and polystyrene fires, and that the unburned hydrocarbons did not represent a significant fraction of the carbon in the upper layer.

Figure 4.9 shows the gas species mass fractions at the front and rear of the compartment as a function of mixture fraction for the iso-propanol fire tests #14, #15 and #30. Figure 4.9a shows all of the transient measurements, and Figure 4.9b and c show the averaged quasi-steady measurements as a function of mixture fraction with and without soot, respectively. From near-stoichiometric conditions ($Z > 0.8$), Y_{CO} increases rapidly, which leads to Y_{CO_2} lower than predicted by the mixture fraction model. The value of Y_{CO} is as high as 7 %. In Figure 4.9b, the measured data are distributed over the range of $0.09 < Z < 0.13$, while those in Figure 4.9c are over the range of $0.12 < Z < 0.15$. Different from above results for other fuels, even the sum of UH and soot does not reach the UH line by the mixture fraction model.

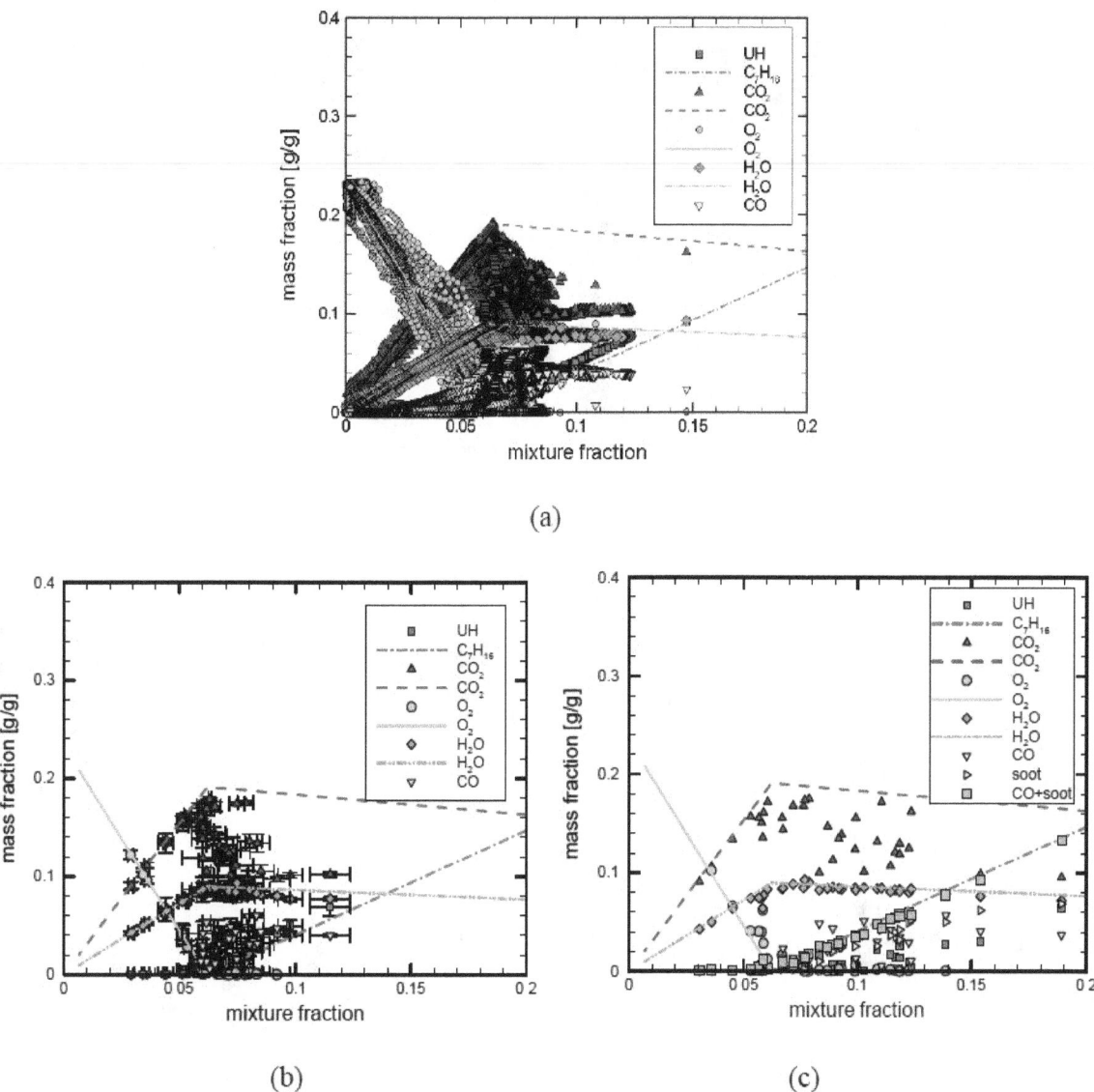

Figure 4.4. Mass fractions of front and rear compartment gas species for the heptane fire tests #5, #9, #12, #13, #19, #22-#26 and #28: (a) transient measurements, (b) time-averaged measurements as a function of mixture fraction without soot, and (c) time-averaged measurements as a function of mixture fraction including soot.

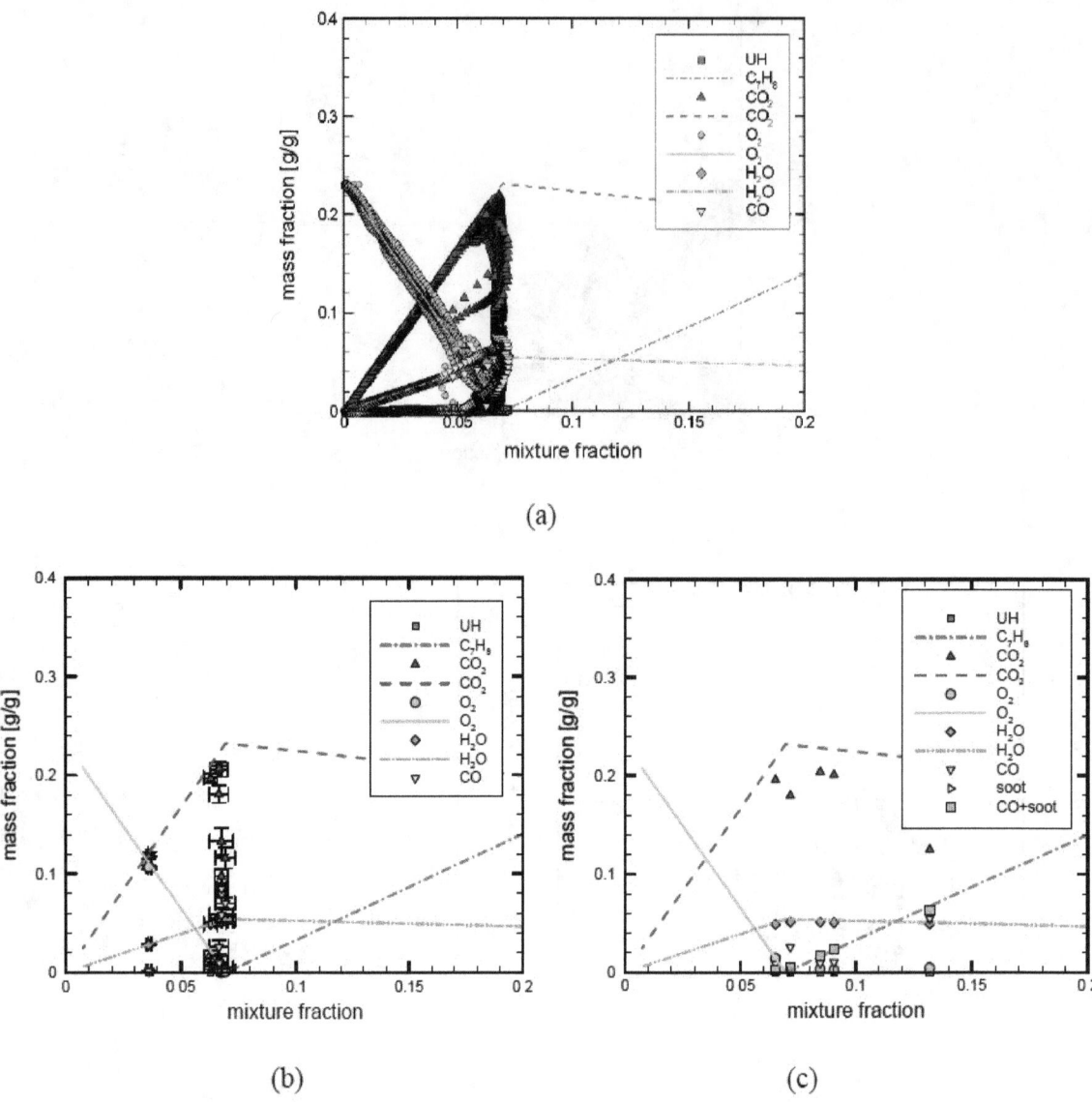

Figure 4.5: Mass fractions of front and rear compartment gas species for the toluene fire tests #20 and #29: (a) transient measurements, (b) time-averaged measurements as a function of mixture fraction without soot, and (c) time-averaged measurements as a function of mixture fraction including soot.

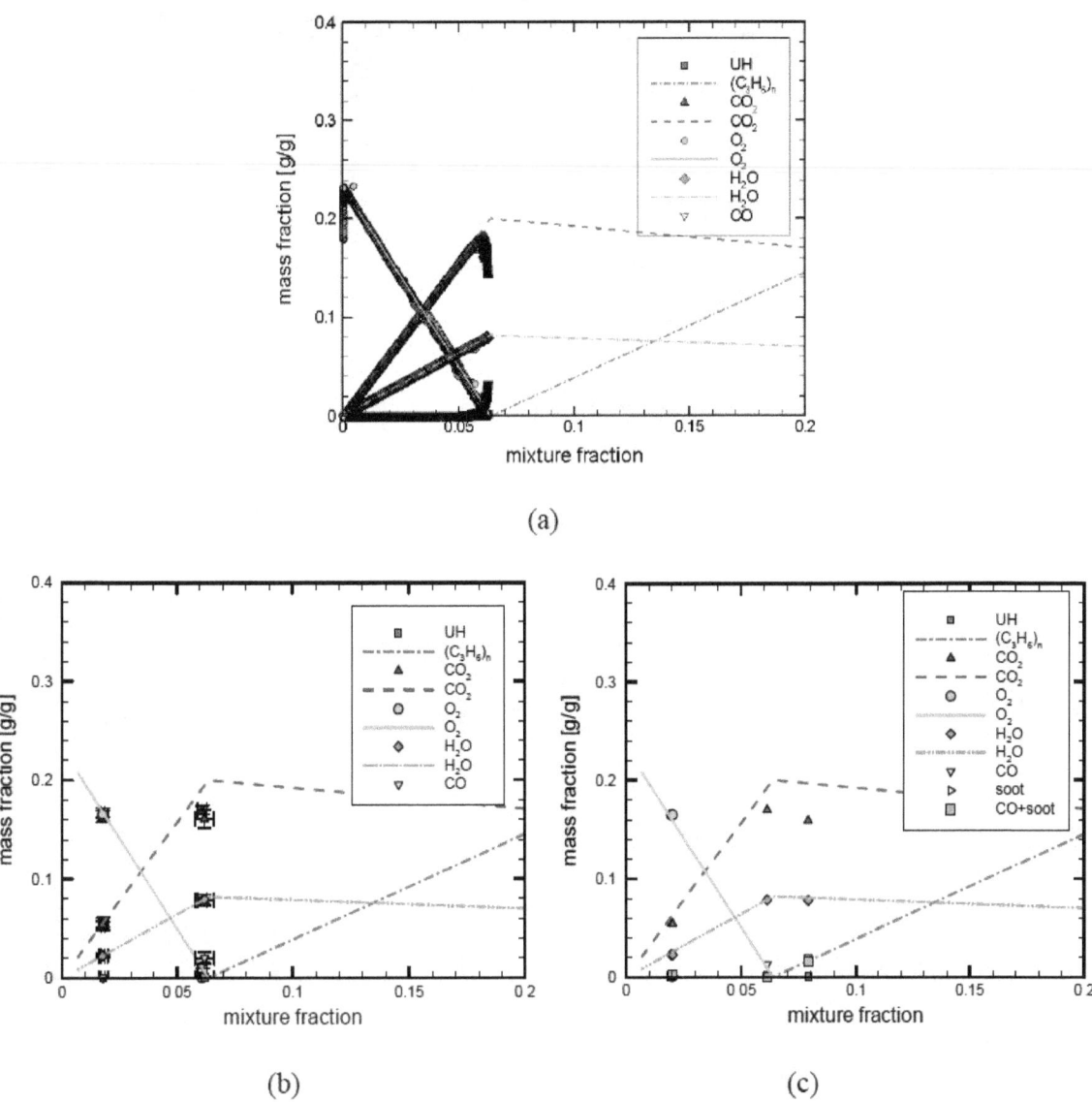

Figure 4.6: Mass fractions of front and rear compartment gas species for the polypropylene fire tests #11 and #18: (a) transient measurements, (b) time-averaged measurements as a function of mixture fraction without soot, and (c) time-averaged measurements as a function of mixture fraction including soot.

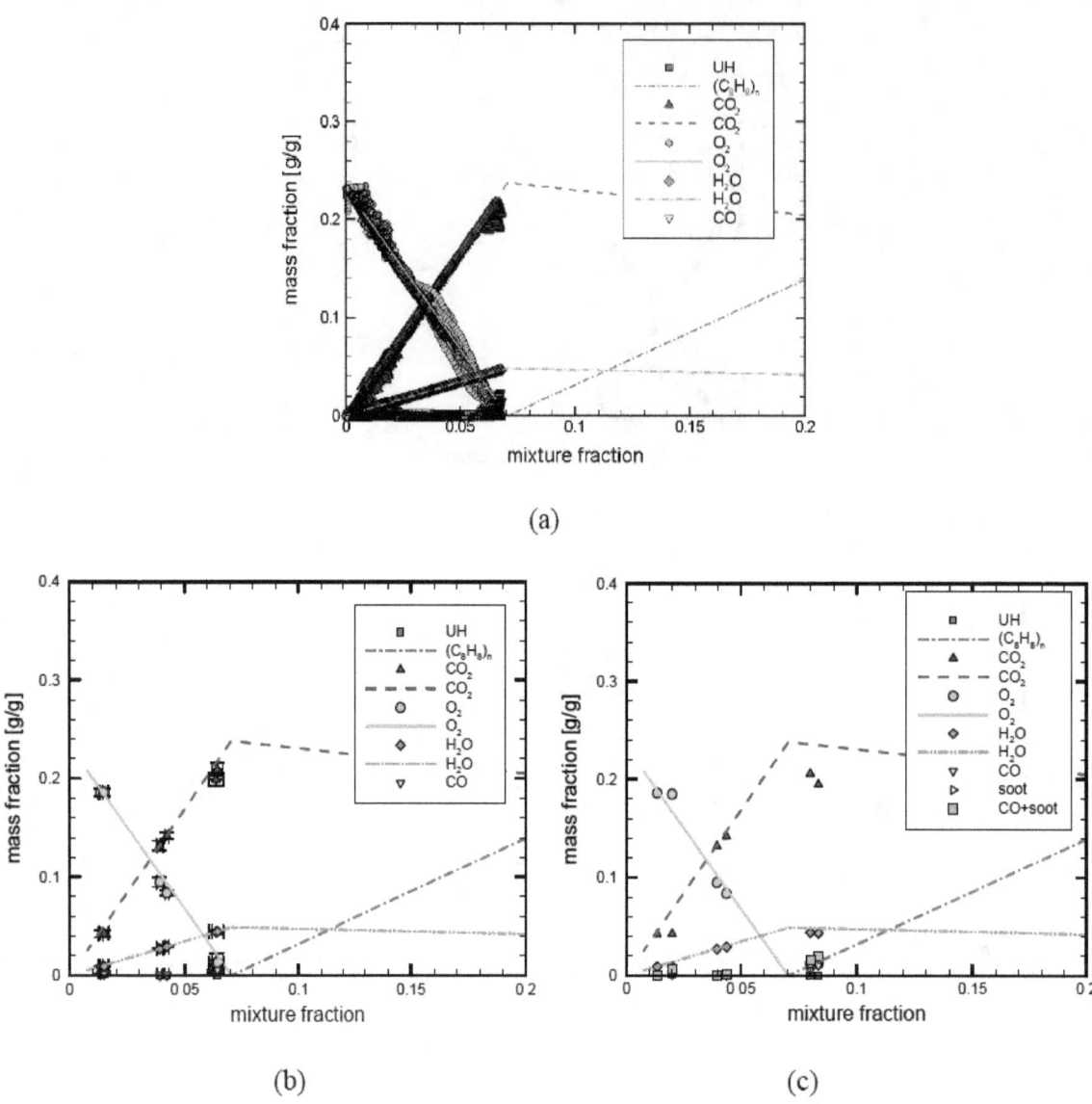

Figure 4.7: Mass fractions of front and rear compartment gas species for the polystyrene fire tests #16, #17 and #21: (a) transient measurements, (b) time-averaged measurements as a function of mixture fraction without soot, and (c) time-averaged measurements as a function of mixture fraction including soot.

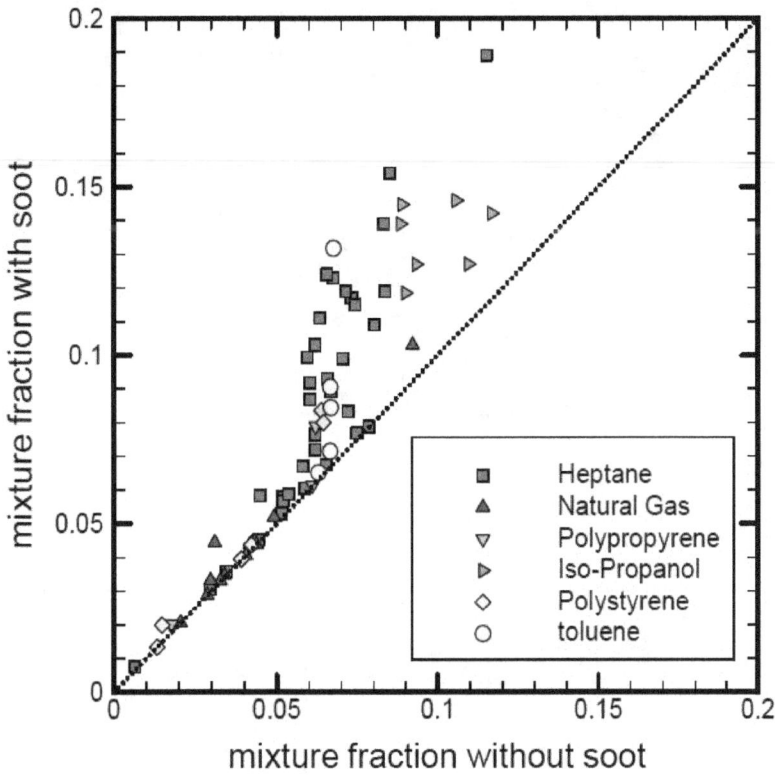

Figure 4.8: Comparison of mixture fraction calculated with and without soot using the time-averaged species measurements when the HRR was quasi-steady.

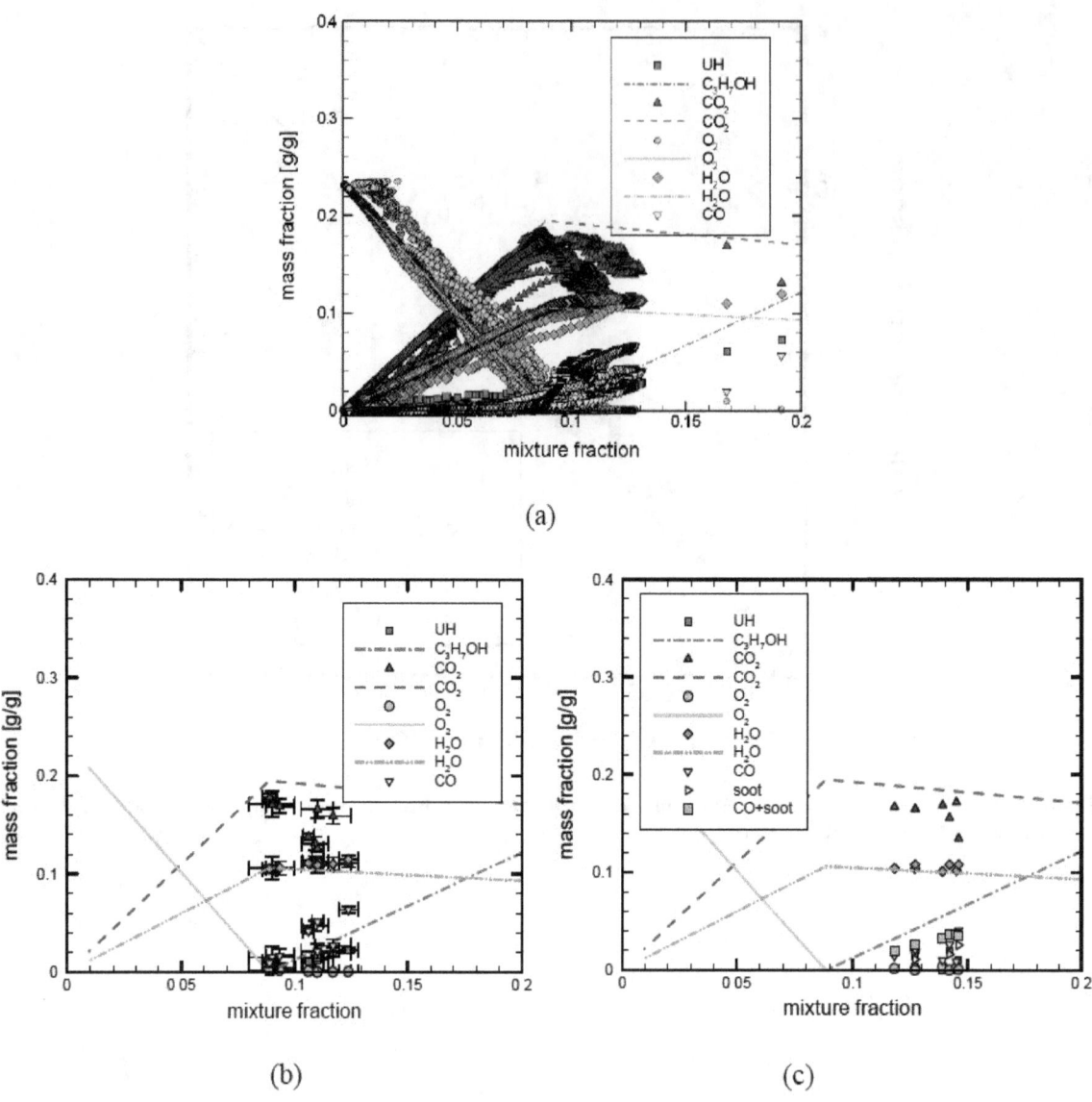

Figure 4.9: Mass fractions of front and rear compartment gas species for the iso-propanol fire tests #14, #15 and #30: (a) transient measurements, (b) time-averaged measurements as a function of mixture fraction without soot, and (c) time-averaged measurements as a function of mixture fraction including soot.

4.2 Post-Compartment Product Yields

The behavior of under-ventilated compartment fires may be divided into burning inside and burning outside of the compartment. Therefore, to investigate the overall fire characteristics of the compartment fire, it is important to consider the product emissions in the exhaust stack including those inside the compartment which flow out. The information in the exhaust stack can be used to estimate the overall combustion efficiency and estimate the fraction of burning occurring within the compartment for complete combustion.

Figure 4.10 presents the pseudo-steady state averages of CO_2 volume fraction, X_{CO_2}, determined from the measurements made in the exhaust stack during the quasi-steady burning periods (indicated in Table 3.3), plotted as a function of heat release rate. Because overall fire behaviors may be affected by doorway size indicating the ventilation condition, only two ventilation conditions, i.e. full doorway (80 cm) and 1/4 doorway (20 cm), are plotted in the figure. In the figure, DF is defined as the Doorway Fraction, that is, the fraction of a full scale ISO 9705, 80 cm doorway. As the heat release rate is increased there is an essentially linear increase of X_{CO_2} for all fuels including natural gas, heptane, iso-propanol, polypropylene, polystyrene, and toluene. Comparing the value of X_{CO_2} between DF=1.00 and 0.25 as a function of heat release rate, there only difference is that the value of X_{CO_2} at DF=0.25 is somewhat larger than at DF=1.0. On the other hand, there are clear differences among the different fuel tested. For example, consider the value of X_{CO_2} as a function of heat release rate, while polypropylene, polystyrene, toluene and iso-propanol show a large X_{CO_2}, natural gas has the smallest value. As the heat release rate is increased, the difference in CO_2 emission among the different fuels increases. Therefore, it can be observed that overall CO_2 emission is affected more by heat release rate and fuel type rather than vent size.

Figure 4.11 presents the pseudo-steady state CO volume fraction, X_{CO}, measurements from the exhaust stack as a function of heat release rate for the same fuels and ventilation conditions as shown in Figure 4.10. For DF=1.00, X_{CO} increases continuously as heat release rate increases, even though X_{CO} is very low compared to that observed in the DF=0.25 case. This may be explained as CO emission rate increasing as heat release rate increases due to insufficient O_2 within the room. However, in the case of DF=0.25, X_{CO} exhibits different behavior compared to the case of DF=1.00. The maximum values of CO emission occur near the heat release rate of 1250kW and then drop off at higher heat release rates. This may be an indication of a transition from burning inside the room to burning outside the room. That is, as the heat release rate continues to increase more CO is produced within the room that is not consumed after it exits the room. However, beyond some critical point there is sufficient unburnt fuel and intermediate combustion products coming out of the room that they are able to support a fire outside of the room which then consumes the remaining CO. This observation may not be conclusive though, since there were relatively few DF=0.25 cases with a heat release rate greater than 1500 kW.

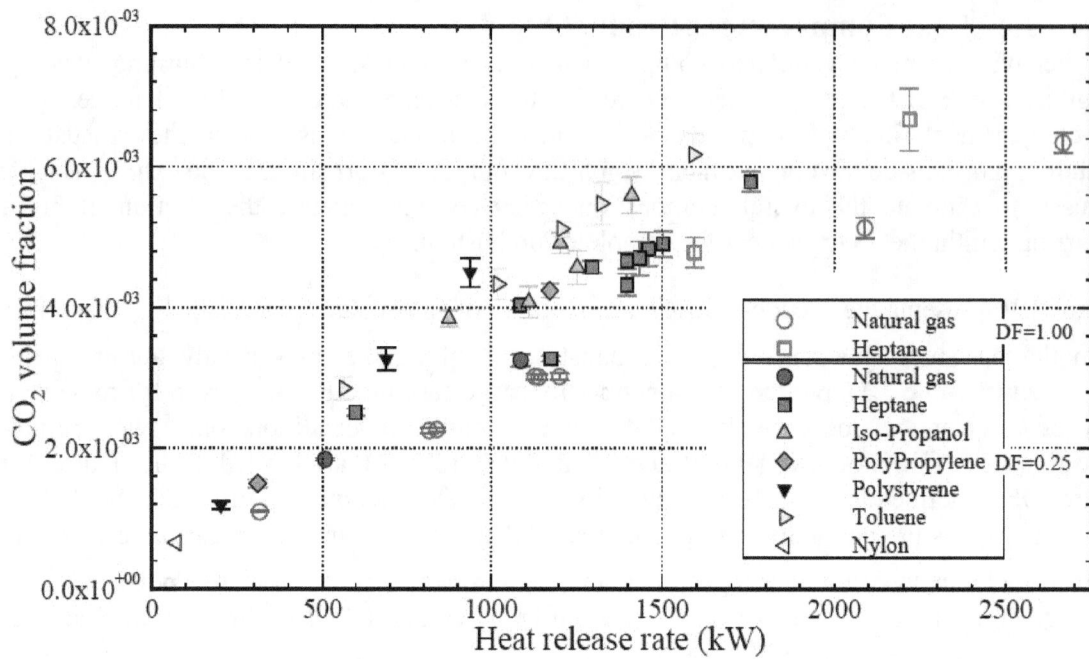

Figure 4.10: The CO_2 volume fraction, X_{CO_2}[2], in the exhaust stack as a function of the fire heat release rate during the periods when the HRR was quasi-steady for each of the fuels tested (DF indicates the doorway fraction of 80 cm).

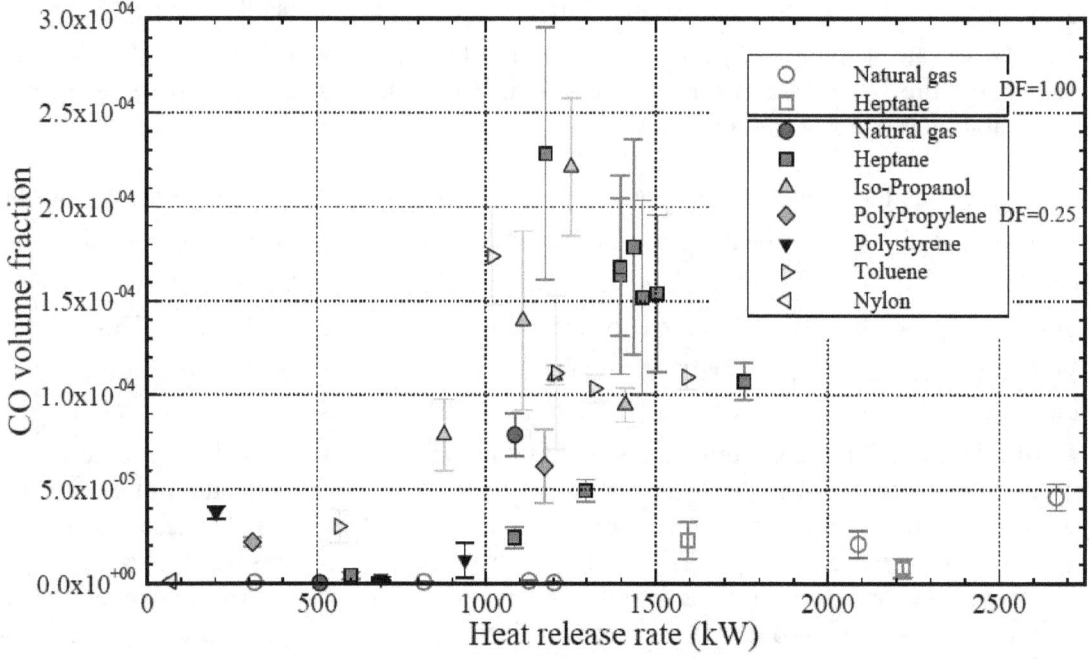

Figure 4.11: The CO volume fraction, X_{CO} 2, in the exhaust stack as a function of the fire heat release rate during the periods when the HRR was quasi-steady for each of the fuels tested.

---

[2] CO and CO_2 volume fractions are measured as absolute values and not adjusted for various hood flow rates that may vary between data points. Adjusting for hood flow rate would shift the slope of the volume fractions but would not affect the qualitative behavior illustrated here.

4.3 Carbon Balance

Compartment measurements show that elemental carbon was primarily distributed among soot and three principal gaseous species (CO_2, CO, and CH_4) in the upper layer of the compartment. Other hydrocarbons were measured in only trace quantities compared to methane. In our previous study [5], the fractional mass-based amount of carbon was newly used to analyze the species compositions. This new parameter has an advantage that the values are bounded from 0 to 1, contrary to the production yields or generation factor (defined below) which has been typically used to present the composition results. Our previous study [5] showed the trends of the fractional mass-based amount of carbon were very similar in appearance to those of the production yields or generation factor, indicating the new parameter is a reasonable way to represent the composition results.

This fractional mass-based amount of carbon that existed in the form of carbon monoxide (F_{CO}) or carbonaceous soot (F_{soot}) is related to the mass fractions of carbon containing species at each measurement location as:

$$F_{soot} = \frac{Y_{soot}}{\frac{12}{16}Y_{CH_4} + \frac{12}{44}Y_{CO_2} + \frac{12}{28}Y_{CO} + Y_{soot}} \; ; \; F_{CO} = \frac{\frac{12}{28}Y_{CO}}{\frac{12}{16}Y_{CH_4} + \frac{12}{44}Y_{CO_2} + \frac{12}{28}Y_{CO} + Y_{soot}} \quad 27$$

In the results presented for the compartment data, the value of X_s, which is a representation of the amount of carbonaceous soot is defined as:

$$X_s = \frac{Y_{soot}}{MW_C} \Big/ \sum \frac{Y_i}{MW_i} \quad 28$$

which comes directly from algebraic manipulation of Eqs. 15 and 16 (in Sec 4.1.1), and the facts that $\sum X_i MW_i$ is a constant, and $\sum X_i = 1$.

Table 4.2 lists F_{soot} and F_{CO} based on averages of the quasi-steady species measurements at the front and rear locations during each of the fires (heptane, natural gas, polypropylene, polystyrene, iso-propanol and toluene). For convenience, the fire heat release rate (HRR), the local equivalence ratio (ϕ) and the ratio (F_{CO}/F_{soot}) are also included in the table. Also, Figure 4.12a and Figure 4.12b show the F_{CO} and F_{soot} as a function of the local equivalence ratio for all fuels. The value of F_{soot} was different for the different fuels, tending to increase with the local equivalence ratio (or mixture faction). The F_{soot} was largest for the toluene fires and the heptane fires (see Table 4.2), reaching a value of over 0.50. This means that in those cases, about half of the carbon exists in the form of soot. The F_{soot} in the other fires was also large, taking on values as large as 0.31 in one of the natural gas fires, 0.23 in one of the polypropylene fires, 0.41 in one of the iso-propanol fires, and 0.26 in one of the polystyrene fires. The value of F_{CO} was as large as 0.003 to 0.30 for the heptane fires, 0.0 to 0.18 for the natural gas fires, 0.02 to 0.13 for the polypropylene fires, 0.05 to 0.2 for the iso-propanol fires, 0.002 to 0.06 for the polystyrene fires, and 0.05 to 0.2 for the toluene fires. In the richest toluene fire (ϕ=1.93), the sum of F_{CO} and F_{soot}

reached to 0.72, indicating that over 70 % of the carbon exists in the form of CO or soot, with relatively little carbon in the form of CO_2 or unburned fuel. The maximum values of the sum of F_{CO} and F_{soot} were 0.67, 0.31, 0.35, 0.49, and 0.29 for the heptane, natural gas, polypropylene, iso-propanol, and polystyrene, respectively. Table 4.2 lists value of F_{CO}/F_{soot}, which depends on fuel type, and physical location. Its value was less than 1.0, except in a few cases of the heptane fires and natural gas fires.

Measurements by Koylu et al [46] and Puri and Santoro [47] showed that there is a linear relation in the emission of soot and CO from buoyant turbulent diffusion flames burning various hydrocarbon fuels (acetylene, propene, etc.). Measurements in the fuel lean (overfire) plume region of hydrocarbon fires showed that the soot and CO generation factors (η_S and η_{CO}) tended to increase with flame residence time, until a near-constant value was reached after long times (compared to the smoke point). Koylu et al [46] reported that the ratio of the CO and soot generation factors for a range of fuel types was such that, $\eta_{CO}/\eta_S = 0.34 \pm 0.09$. The generation rate was defined as the mass of soot (or gas species) produced per unit mass of fuel carbon consumed. This is slightly different than the soot (or gas species) yield (y_{CO} and y_S), which is based on the mass of all elements (not just carbon) in the fuel stream. The ratios of the yields and the generation rates, however, are equal, and their value can be determined at any location from the ratio of the mass fractions of CO and soot:

$$\eta_{CO}/\eta_{soot} = y_{CO}/y_{soot} = Y_{CO}/Y_{soot} = (7/3) F_{CO}/F_{soot} \qquad 29$$

The constant value (7/3) in Eq. 29 is the ratio of the total CO mass to the mass of carbon.

Table 4.3 lists y_{co}, y_{soot}, and the ratio y_{co}/y_{soot} based on the time-averaged species measurements at the front and rear compartment locations when the heat release rate was quasi-steady during each of the fires (heptane, natural gas, polypropylene, iso-propanol, polystyrene and toluene). The fire heat release rate (HRR) and the local equivalence ratio are also listed. Much of the same data was used as in Table 4.2. Figure 4.13a and Figure 4.13b also show the yield of CO and soot as a function of the local equivalence ratio for all fuels (heptane, natural gas, polypropylene, polystyrene, iso-propanol and toluene). Figure 4.13 is analogous to Figure 4.12, with the parameters y_{CO} and y_{soot} considered in lieu of F_{CO} and F_{soot}. The trends and values of the data shown in the graphs are very similar in appearance, consistent with the data presented in Table 4.2, and this shows that F_{CO} and F_{soot} is a reasonable way to represent the composition results.

Figure 4.14 shows the ratio of the CO yield to the soot yield as a function of the local equivalence ratio for the same quasi-steady data shown in Figure 4.13 for the heptane, natural gas, polypropylene, iso-propanol, and toluene fires. Figure 4.15 shows the CO yield as a function of the soot yield for the same quasi-steady data for the heptane, natural gas, polypropylene, iso-propanol, and toluene fires. Shown is a line representing the results of Koylu et al [46]. Koylu reported about 30 % scatter in the ratio of the yields of CO to soot, which is considerably smaller than that seen in the figure. Nevertheless, more data are needed to examine this relationship in the upper layer of compartment fires. It is interesting to note that Tewarson et al [48] reported that the ratio of the CO and soot generation efficiencies from small fires burning polymers varied, depending on the exact fuel type and the amount of ventilation.

Table 4.2: Average fractional soot, CO and CO/soot ratio at the front and rear compartment measurement locations.

Fuel	HRR [kW]	Rear				Front			
		ϕ_{local}	F_{CO}	F_{soot}	F_{CO}/F_{soot}	ϕ_{local}	F_{CO}	F_{soot}	F_{CO}/F_{soot}
Heptane	271	-	-	-	-	0.473	0.003	0.033	0.088
	414	-	-	-	-	0.895	0.003	0.240	0.013
	538	0.968	0.030	0.033	0.906	1.729	0.026	0.453	0.057
	582	1.208	0.053	0.201	0.265	1.954	0.044	0.500	0.087
	600	-	-	-	-	0.111	0.011	0.149	0.072
	719	-	-	-	-	1.905	0.149	0.424	0.351
	923	-	-	-	-	2.557	0.135	0.480	0.282
	991	-	-	-	-	0.558	0.006	0.036	0.159
	1086	0.926	0.017	0.088	0.189	1.356	0.121	0.323	0.376
	1093	-	-	-	-	3.380	0.100	0.432	0.231
	1297	-	-	-	-	1.599	0.254	0.420	0.604
	1381	0.710	0.010	0.019	0.526	0.839	0.005	0.030	0.178
	1397	1.367	0.295	0.144	2.053	1.860	0.187	0.377	0.496
	1397	1.475	0.135	0.310	0.436	1.792	0.140	0.282	0.498
	1435	1.951	0.121	0.478	0.254	1.977	0.130	0.321	0.405
	1460	1.421	0.249	0.266	0.939	1.589	0.186	0.306	0.608
	1503	-	-	-	-	1.892	0.152	0.401	0.378
	1594	1.056	0.105	0.141	0.743	1.137	0.094	0.148	0.638
	1756	-	-	-	-	2.295	0.210	0.430	0.488
	1762	-	-	-	-	1.427	0.157	0.363	0.433
	1796	-	-	-	-	1.538	0.067	0.422	0.158
	2218	0.908	0.041	0.110	0.370	0.891	0.014	0.087	0.162
Natural Gas	507	0.572	0.000	0.110	0.000	0.732	0.000	0.309	0.000
	1087	-	-	-	-	1.901	0.176	0.111	1.584
	1140	0.786	0.001	0.005	0.133	0.612	0.004	0.022	0.175
	1203	0.508	0.001	0.006	0.160	0.584	0.004	0.013	0.307
	2089	0.776	0.006	0.050	0.113	0.924	0.030	0.057	0.527
Polypropylene	311	0.288	0.018	0.083	0.213	0.298	0.018	0.125	0.142
	1174	-	-	-	-	1.216	0.126	0.227	0.555
Iso-Propanol	875	1.527	0.052	0.384	0.136	1.592	0.045	0.408	0.110
	1111	1.312	0.089	0.255	0.348	1.473	0.118	0.147	0.797
	1205	-	-	-	-	1.664	0.200	0.294	0.680
	1254	1.414	0.091	0.280	0.324	1.665	0.133	0.192	0.691
Polystyrene	203	-	-	-	-	0.267	0.032	0.257	0.124
	690	0.542	0.002	0.010	0.231	0.601	0.003	0.037	0.076
	939	1.177	0.058	0.250	0.230	1.127	0.031	0.207	0.151
Toluene	1020	-	-	-	-	1.212	0.053	0.223	0.235
	1209	-	-	-	-	1.299	0.053	0.281	0.189
	1323	-	-	-	-	1.931	0.196	0.519	0.377

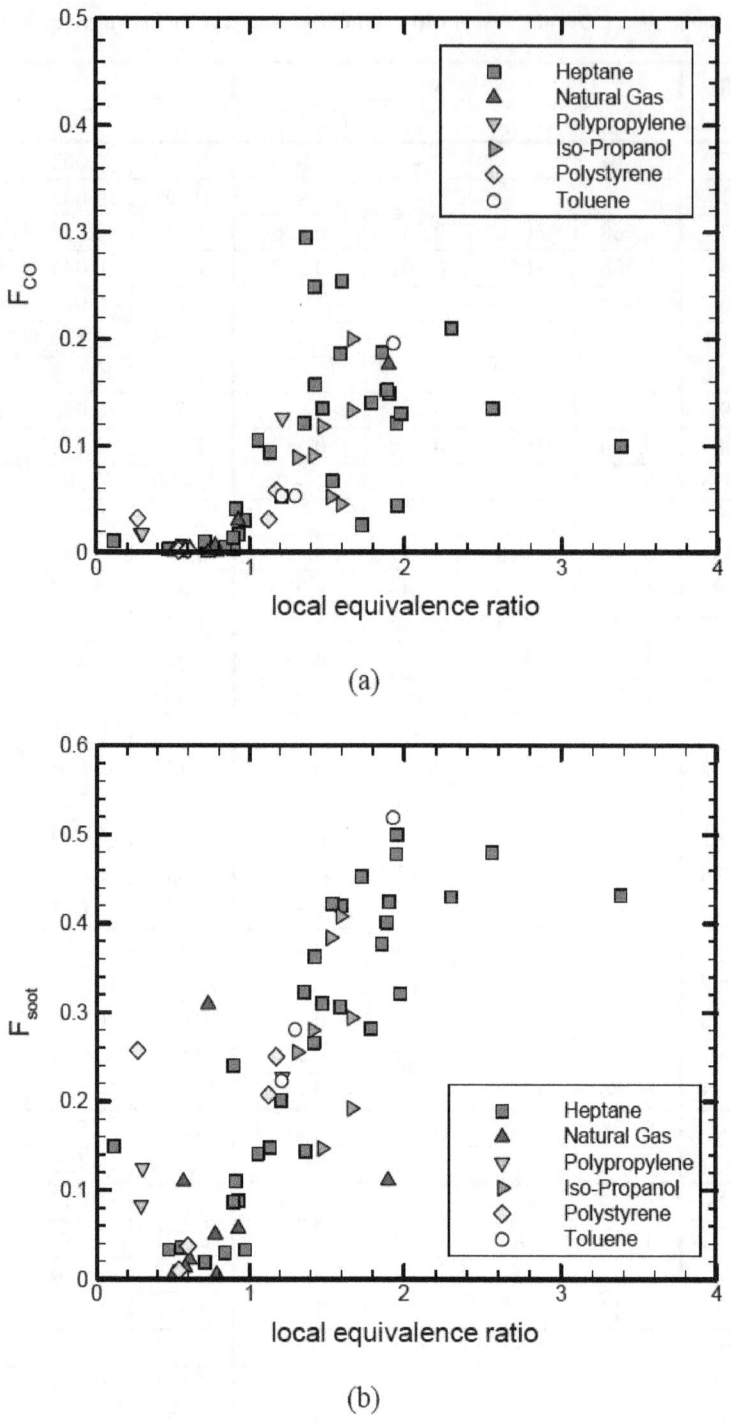

Figure 4.12: The values of F_{CO} and F_{soot} as a function of the local equivalence ratio for the time averaged measurements during the period when the HRR was quasi-steady.

Table 4.3: Average yields of soot, CO and CO/soot ratio at the front and rear compartment measurement locations.

Fuel	HRR [kW]	Rear				Front			
		ϕ_{local}	y_{CO}	y_{soot}	y_{CO}/y_{soot}	ϕ_{local}	y_{CO}	y_{soot}	y_{CO}/y_{soot}
Heptane	271	-	-	-	-	0.473	0.006	0.028	0.206
	414	-	-	-	-	0.895	0.006	0.202	0.029
	538	0.968	0.058	0.027	2.113	1.729	0.050	0.381	0.132
	582	1.208	0.105	0.169	0.618	1.954	0.086	0.421	0.204
	600	-	-	-	-	0.111	0.022	0.129	0.168
	719	-	-	-	-	1.905	0.293	0.357	0.820
	923	-	-	-	-	2.557	0.266	0.404	0.659
	991	-	-	-	-	0.558	0.011	0.031	0.371
	1086	0.926	0.033	0.074	0.442	1.356	0.239	0.272	0.877
	1093	-	-	-	-	3.380	0.196	0.364	0.539
	1297	-	-	-	-	1.599	0.498	0.354	1.408
	1381	0.710	0.020	0.016	1.228	0.839	0.011	0.026	0.416
	1397	1.367	0.580	0.121	4.788	1.860	0.367	0.317	1.157
	1397	1.475	0.266	0.261	1.017	1.792	0.276	0.237	1.162
	1435	1.951	0.239	0.403	0.592	1.977	0.256	0.270	0.945
	1460	1.421	0.490	0.224	2.190	1.589	0.366	0.258	1.419
	1503	-	-	-	-	1.892	0.298	0.338	0.883
	1594	1.056	0.205	0.119	1.732	1.137	0.186	0.125	1.487
	1756	-	-	-	-	2.295	0.412	0.362	1.137
	1762	-	-	-	-	1.427	0.309	0.306	1.010
	1796	-	-	-	-	1.538	0.131	0.356	0.368
	2218	0.908	0.080	0.093	0.862	0.891	0.028	0.073	0.379
Natural Gas	507	0.572	0.000	0.083	0.000	0.732	0.000	0.233	0.000
	1087	-	-	-	-	1.901	0.308	0.083	3.695
	1140	0.786	0.001	0.003	0.311	0.612	0.007	0.017	0.407
	1203	0.508	0.002	0.004	0.374	0.584	0.007	0.010	0.715
	2089	0.776	0.010	0.038	0.264	0.924	0.053	0.043	1.229
Polypropylene	311	0.288	0.036	0.072	0.496	0.298	0.036	0.108	0.331
	1174	-	-	-	-	1.216	0.253	0.195	1.294
Iso-Propanol	875	1.527	0.100	0.315	0.317	1.592	0.086	0.334	0.258
	1111	1.312	0.170	0.209	0.812	1.473	0.225	0.121	1.860
	1205	-	-	-	-	1.664	0.382	0.241	1.586
	1254	1.414	0.174	0.230	0.757	1.665	0.255	0.158	1.612
Polystyrene	203	-	-	-	-	0.267	0.069	0.240	0.289
	690	0.542	0.005	0.009	0.539	0.601	0.006	0.034	0.178
	939	1.177	0.124	0.232	0.536	1.127	0.067	0.192	0.352
Toluene	1020	-	-	-	-	1.212	0.112	0.204	0.549
	1209	-	-	-	-	1.299	0.114	0.257	0.442
	1323	-	-	-	-	1.931	0.417	0.475	0.879

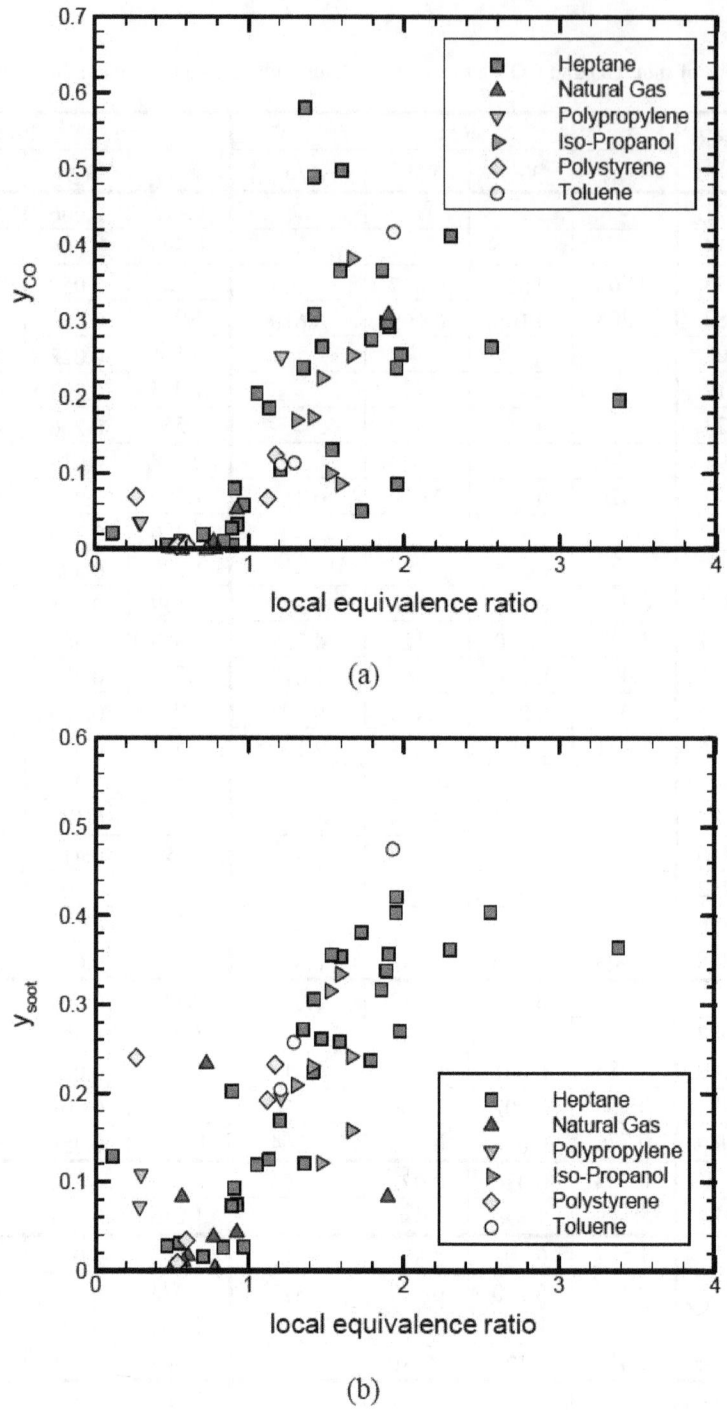

Figure 4.13: The CO and soot yields as a function of the local equivalence ratio for the time averaged measurements during the period when the HRR was quasi-steady.

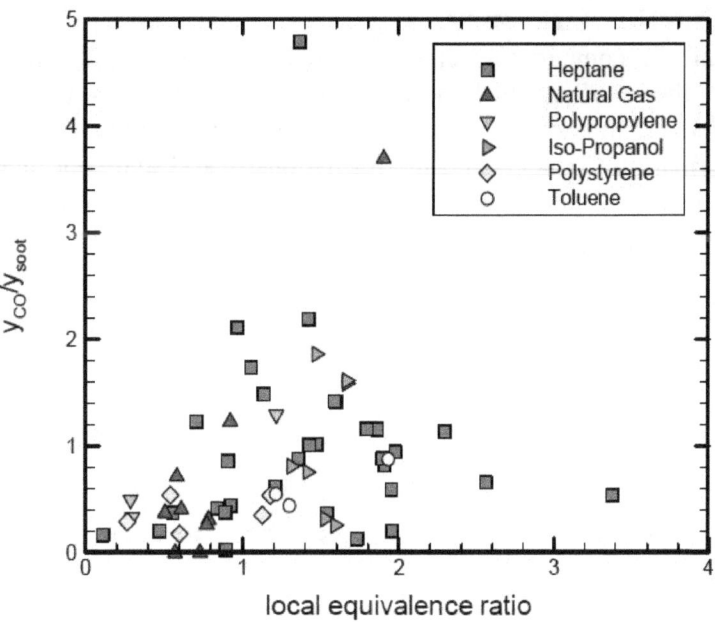

Figure 4.14: The ratio of the CO to soot yield as a function of the local equivalence ratio during the period when the HRR was quasi-steady.

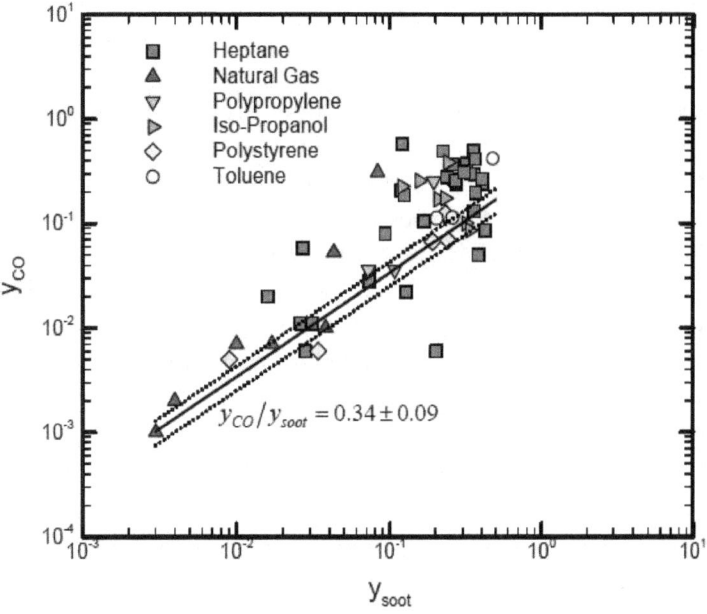

Figure 4.15: The CO yield as a function of the soot yield during the period when the HRR was quasi-steady. Also shown is a line representing the results of Koylu [2].

4.4 Combustion Efficiency

To better understand the compartment chemistry, it is of interest to determine the combustion efficiency both in the exhaust stack and at various locations in the upper layer of the compartment. Moreover, the accurate prediction of burning fraction inside compartment may provide useful information to understand the formations of CO and soot including detailed fire dynamics. The combustion efficiency (χ_a) is a global representation of the fractional amount of heat released by the fire as compared to complete combustion. It is defined as:

$$\chi_a = \frac{\Delta H_c}{\Delta H_{c,ideal}} \qquad 30$$

where $\Delta H_{c,ideal}$ is the net heat of complete combustion based on the conversion of all carbon and hydrogen in the fuel to CO_2 and H_2O (assumed to remain in the vapor phase) and ΔH_c is the net heat of combustion, which is the actual heat released in a chemical reaction. The value of χ_a is bounded by 0 % to 100 %.

The local combustion efficiency at the front and rear locations in the upper layer of the compartment was calculated based on the difference in the heat of formation of species measured using gas analyzer. The combustion efficiency at the exhaust stack was determined as the ratio of the measured heat release rate to the measured mass delivery rate. A summary of averaged steady-state results of combustion efficiency in the exhaust stack is shown in Table 4.4.

Figure 4.16 and Figure 4.17 show the combustion efficiency in the exhaust stack using measurements made during the steady state burning periods (indicated in Table 3.5) as a function of the ideal heat release rate for the condition of full doorway (DF=1.00) and 1/4 doorway size (DF=0.25), respectively. Figure 4.16 presents the results for natural gas and heptane fuels corresponding to ISONG1~3 and ISOHept4~5. Ideal heat release rate was calculated based on the measured mass delivery rate and ideal heat of combustion. Some natural gas cases have average combustion efficiencies above 100 % but still within the uncertainty of the hood, heptane has a maximum combustion efficiency of approximately 90 %. The combustion efficiencies above 100 % are a result of the uncertainty associated with the heat release rate measurement. As mentioned in section 2.2 heat release rate measurements have the combined expanded relative uncertainty of 14 %. For the DF=0.25 (1/4 door) case the combustion efficiency decreases as the ideal heat release rate increases and there is similar behavior among all the fuels tested.

Figure 4.18 presents the combustion efficiency in the exhaust stack as a function of the ideal heat release rate for several doorway sizes in heptane spray burner fires. The tests plotted are ISOHept22~25. In this figure, all ventilation conditions show that the combustion efficiency decreases as ideal heat release rate increases as illustrated in Figure 4.17. The combustion efficiency significantly decreases as the doorway width changes from DF=0.500 (1/2 doorway) to DF=0.125 (1/8 doorway).

Figure 4.19 presents the local combustion efficiency at the front and rear sample locations as a function of the ideal heat release rate for the same heptane spray tests with various doorway sizes plotted in Figure 4.18. Comparing Figure 4.18 and Figure 4.19, the combustion efficiency inside compartment of DF=0.500 and 0.250 is higher than that of exhaust stack. It may be attributed to absence of soot yield in the process of calculating the heat of combustion. In the DF=0.125 case, the local combustion efficiency sharply decreases as ideal heat release rate increases while the global combustion efficiency increases. For this situation more of the fuel is burning outside of the compartment while less is burning inside. As a result, the combustion efficiency in the exhaust stack is somewhat larger than that inside the compartment.

Figure 4.20 compares the burning rates inside and outside of the compartment by showing the burning fraction inside compartment and the ratio of combustion efficiency between inside and outside the compartment as a function of ideal heat release rate for the condition of DF=0.125 in heptane spray fires. Averaged combustion efficiency of front and rear sample location was used to represent the combustion efficiency inner compartment. In this figure, for heat release rate less than 750 kW, most of the burning occurs inside the compartment. As the ideal heat release increases from 750 kW to 1500 kW, the portion of burning outside the compartment significantly increases from 0 % to 60 % of the total.

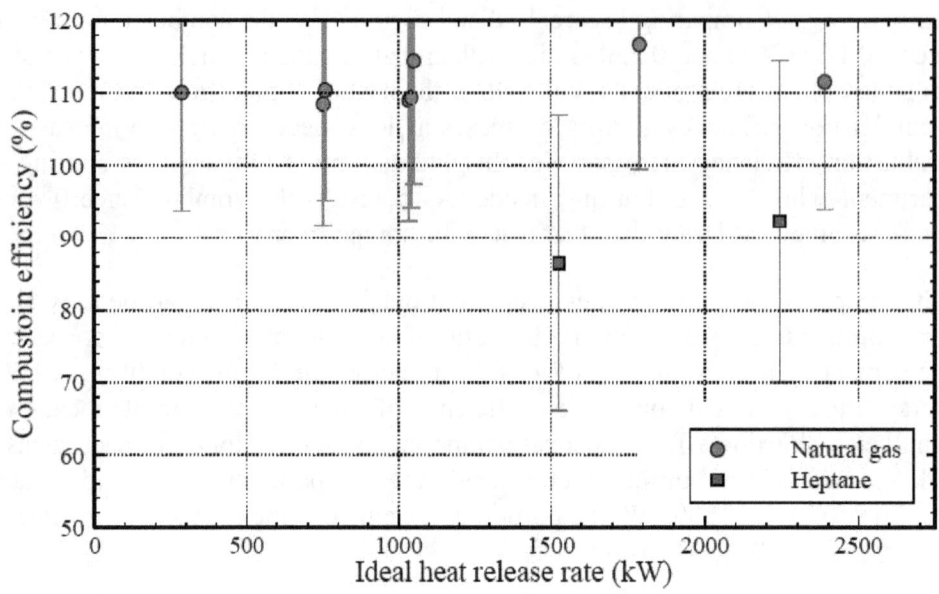

Figure 4.16: The combustion efficiency in the exhaust stack as a function of the ideal heat release rate for natural gas and heptane fuels under the condition of full doorway size (DF=1.0). The indicated uncertainty includes the 14% uncertainty of the calorimetry used to make the measurements.

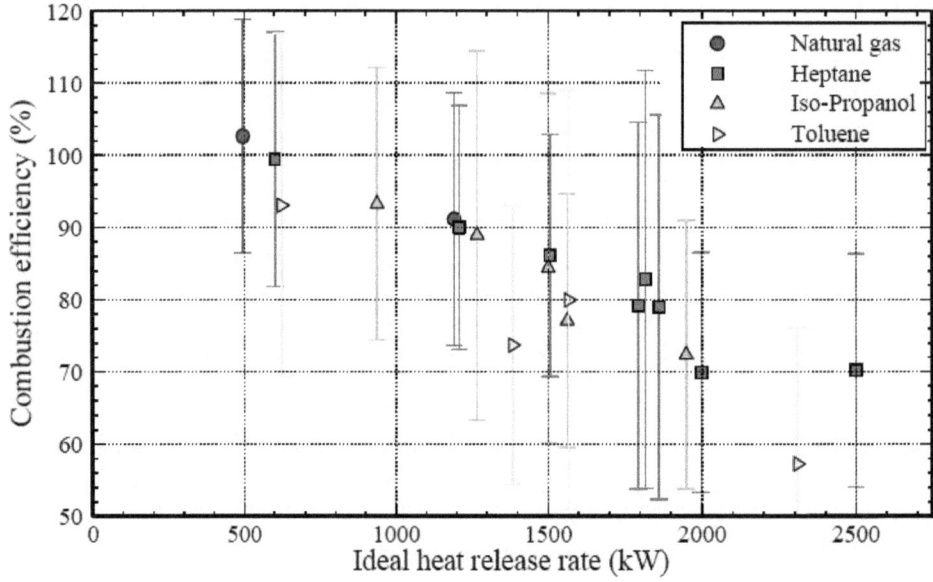

Figure 4.17: The combustion efficiency in the exhaust stack as a function of the ideal heat release rate for various fuels under the condition of DF=0.25. The indicated uncertainty includes the 14% uncertainty of the calorimetry used to make the measurements.

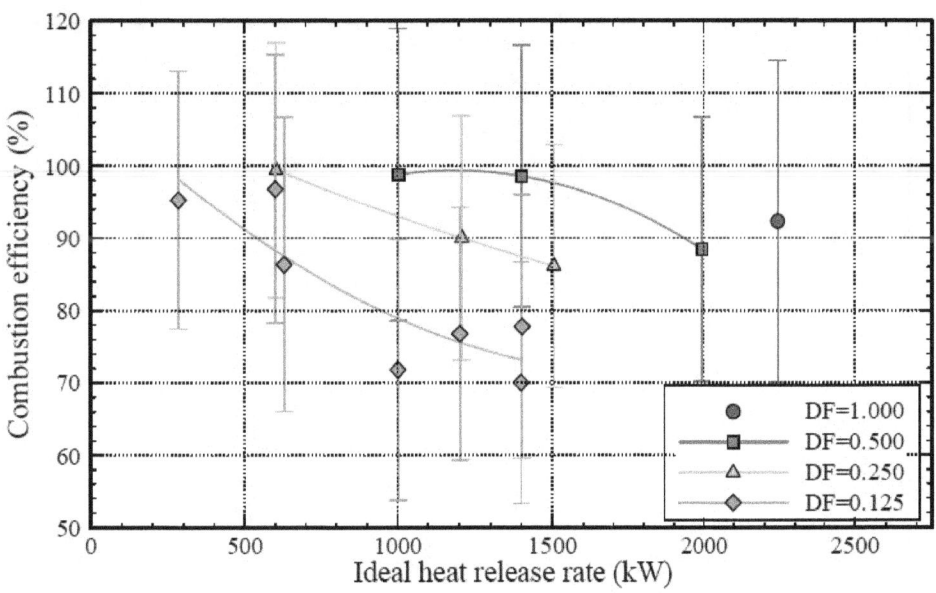

Figure 4.18: The combustion efficiency in the exhaust stack as a function of the ideal heat release rate for various doorway sizes in heptane fires. The curve fit lines are for illustrative purposes only.

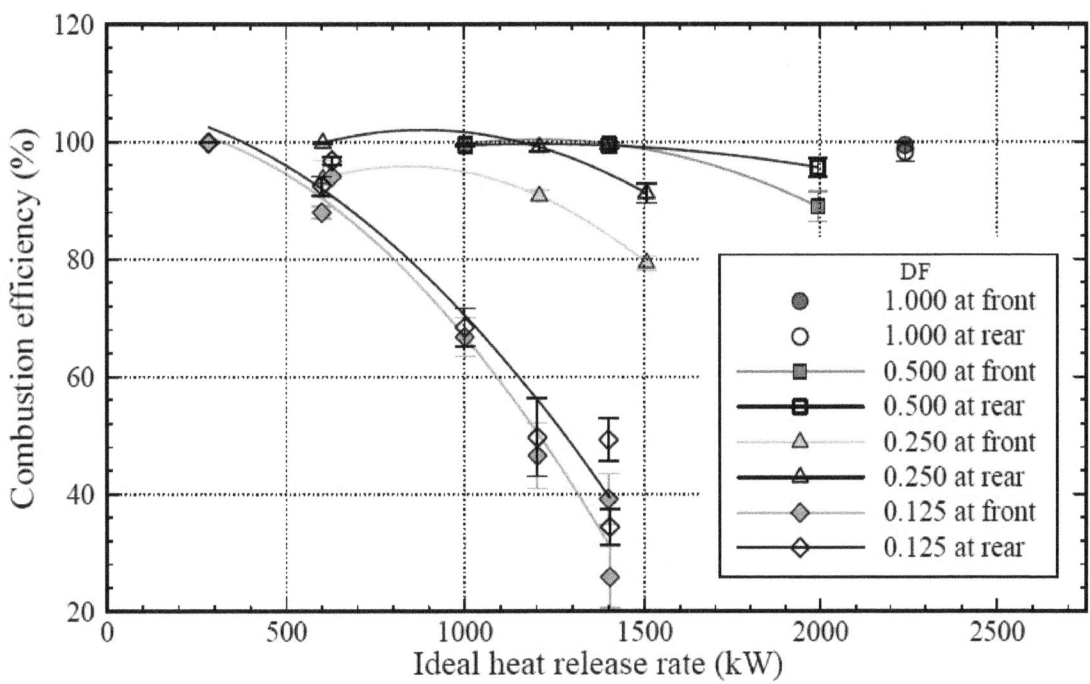

Figure 4.19: The local combustion efficiency at the front and rear sample locations as a function of the ideal heat release rate for various doorway sizes in heptane fires. The curve fit lines are for illustrative purposes only.

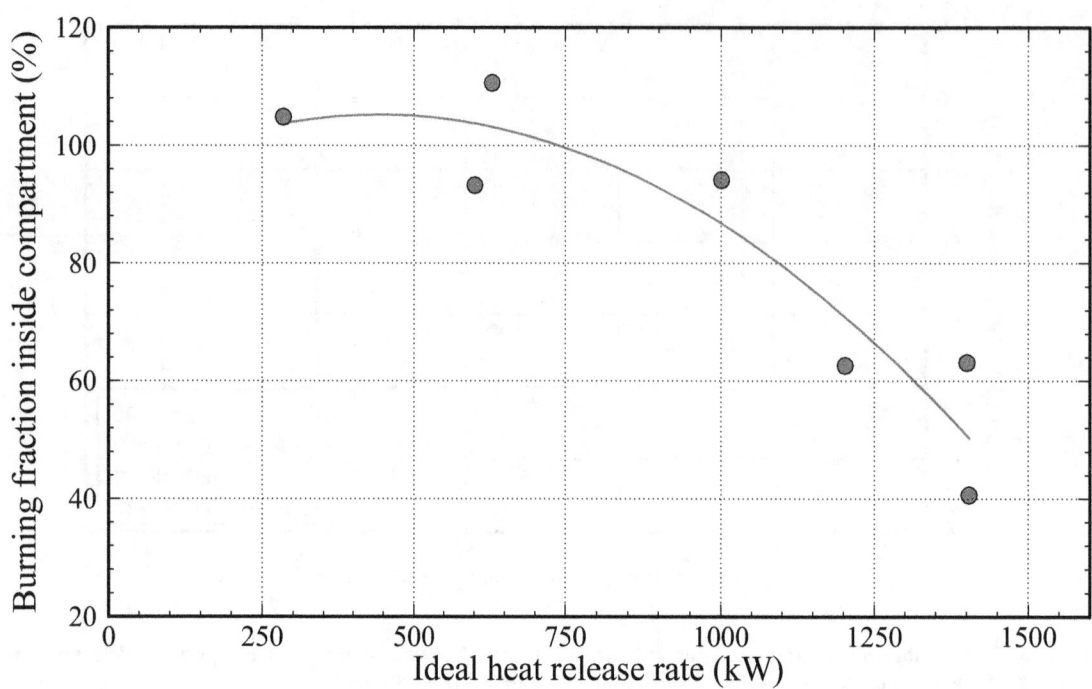

Figure 4.20: The burning fraction inside compartment as a function of ideal heat release rate under the condition of DF=0.125 in heptane fires. The curve fit line is for illustrative purposes only.

Table 4.4: Summary of averaged steady-state results of combustion efficiency in the exhaust stack. The steady state periods here are the same used for all steady state measurements and are listed in Table 3.5. The uncertainty, U, indicated here only reflects the statistical variation

Test No.	Fuel	HRR ideal (Kw)		Combustion Efficiency (%)	
		Mean	U	Mean	U
1	Natural Gas	287.7	2.8	110.05	2.37
		752.1	5.6	108.46	2.78
		1034.8	2.7	109.05	2.79
2	Natural Gas	759.1	0.9	110.38	1.22
		1042.4	4.0	109.38	1.27
3	Natural Gas	1050.9	3.0	114.45	2.94
		1789.6	5.5	116.73	3.25
		2390.3	8.1	111.52	3.62
4	Heptane	2243.5	590.6	92.29	8.24
5	Heptane	1526.5	716.0	86.52	6.44
8	Heptane	1164.6	354.9	114.11	45.49
9	Heptane	1816.8	315.6	82.83	14.93
12	Heptane	1794.0	237.0	79.20	11.34
13	Heptane	1861.9	309.7	78.99	12.63
14	Iso-Propanol	1500.3	149.9	84.40	10.19
15	Iso-Propanol	1268.7	174.0	88.94	11.58
20	Toluene	1566.9	311.9	79.96	15.03
22	Heptane	603.6	3.3	99.44	3.59
		1209.2	38.4	90.00	2.85
		1506.0	1.4	86.13	2.78
23	Heptane	284.4	1.9	95.25	3.83
		629.6	42.3	86.34	6.32
		1001.1	1.1	71.80	4.07
		1400.8	1.9	70.00	2.71
24	Heptane	321.8	55.8	133.15	13.40
		601.0	1.0	96.78	4.54
		1202.7	1.5	76.79	3.43
		1404.4	1.2	77.79	4.24
25	Heptane	1003.3	0.9	98.80	6.16
		1403.4	18.3	98.56	4.10
		1993.5	3.6	88.42	4.23
26	Heptane	1998.8	0.9	88.28	2.59
		2527.6	129.3	82.11	5.12
28	Heptane	1997.9	2.7	69.91	2.57
		2500.1	1.5	70.23	2.14
30	Iso-Propanol	937.4	0.6	93.30	4.89
		1563.2	1.0	77.11	3.55
		1948.9	55.8	72.39	4.61
29	Toluene	623.7	104.5	93.07	9.84
		1383.7	1.3	73.74	5.28
		2309.0	1.6	57.29	4.89
		2890.0	32.4	54.99	3.40
32	Natural Gas	494.4	0.4	102.64	2.18
		1192.3	6.8	91.15	3.49

5 SCALING FROM RSE TO FSE

The applicability of the experimental results reported here to other compartment fire scenarios, including the reduced scale enclosure (RSE) experiments [5], can be considered in terms of a number of normalized parameters traditionally used in fire modeling applications. Use of normalized parameters facilitates comparison of results from scenarios of different scales by normalizing key physical characteristics of the scenario. A number of different forms of scaling may be considered, depending on the fire phenomena of interest [3, 49, 50]. Table 5.1 lists three normalized parameters that may be used to compare fire scenarios with the experiments reported here. The ranges of values for the normalized parameters examined in this study are listed in the table. The table is intended to provide guidance when evaluating the applicability of the data set reported here. For any given fire scenario, more than one normalized parameter may be necessary for determining applicability of the validation results, depending on the parameters of interest, as will be shown here. In this sense, the Table should be considered illustrative, not exhaustive.

The most important parameter of any fire experiment is the heat release rate, as its magnitude drives changes in the thermal environment of the compartment or space of interest. A Froude number expression which normalizes heat release rate to the diameter of the fire, D, is the first entry in Table 5.1, commonly known as Q_d^*, where $\dot{Q}$ is the heat release rate (kW), ρ_∞ is the ambient air density (kg/m³), T_∞ is the ambient air temperature (K), c_p is the specific heat of air (kJ/kg-K), and g is the acceleration of gravity (m/s²). A large value of Q_d^* represents a fire with a relatively large value of energy output power compared to its physical diameter or a large fire plume relative to the burner diameter, like an oil well blowout fire. A low value of Q_d^* represents a fire with a relatively small value of energy output compared to its diameter or a small plume relative to the burner diameter, like a smoldering fire. Many typical accidental fire scenarios have Q_d^* values on the order of 1 which indicates the flow is buoyancy driven as in the case of a pool fire, rather than momentum driven as in the case of a jet flame. The physical diameter of a realistic fire may not be well-defined and may not actually matter when assessing the "size" of a fire. Instead, a characteristic diameter, D^* is considered in the definition of Q_d^* as noted in the table. The range of values of Q_d^* varied as a function of fuel type. In this study, Q_d^* took on values as small as 0.17 and as large as 12.74, varying by 2 orders of magnitude, as seen in Table 5.1. In the context of these experiments the larger values of Q_d^* indicate that the room is likely under-ventilated, having a very large fire within the room, and the smaller values of Q_d^* indicate that the room is likely well ventilated, having a small fire within the room. Figure 5.1 illustrates the two extremes of this parameter contrasting the ISONylon10 fire with a $Q_d^* = 0.17$ where you can see into the room through the door and the ISOPropD14 fire with a Q_d^* of 12.74 where much of the fire is burning outside of the room because the heat release rate is so large relative to the burner sizes.

Table 5.1: List of Normalized Scaling Parameters for Compartment Fires and the Range of Values Examined in this Study.

Parameter	Normalized Representation	Range of Values			Fuel Type
Heat Release Rate [3, 49]	$Q_d^* = \dfrac{\dot{Q}}{\rho_\infty c_p T_\infty \sqrt{gD}D^2}$	0.36	to	12.74	Natural Gas
		0.85	to	10.19	Heptane
		2.12	to	12.74	Toluene
		2.12	to	12.74	Iso-Propanol
		0.71	to	3.93	Polystyrene
		1.07	to	4.29	Polypropylene
		0.17	to	0.68	Nylon
Ventilation [9, 17, 51]	$\phi = \dfrac{\dot{m}_F/\dot{m}_{O_2}}{r}$, where $\dot{m}_{O_2} = \dfrac{0.23}{2} A_o \sqrt{h_o}$	0.13	to	7.90	Natural Gas
		0.13	to	12.48	Heptane
		1.27	to	7.63	Toluene
		1.26	to	7.53	Iso-Propanol
		1.02	to	5.61	Polystyrene
		1.57	to	6.27	Polypropylene
		0.10	to	0.41	Nylon
Compartment height [3, 49]	$\dfrac{H}{D^*}$, where $D^* = \left(\dfrac{\dot{Q}}{\rho_\infty c_p T_\infty \sqrt{g}}\right)^{2/5}$	2.17	to	6.42	Natural Gas
		2.38	to	6.42	Heptane
		2.17	to	4.45	Toluene
		2.17	to	4.45	Iso-Propanol
		2.46	to	4.87	Polystyrene
		2.38	to	4.14	Polypropylene
		7.02	to	12.23	Nylon

The second entry in the table is the global equivalence ratio (ϕ), which is associated with the overall fire-induced ventilation and compartment stoichiometry. An estimate of the maximum achievable steady-state oxygen supply is given by: $\dot{m}_{O_2} = \dfrac{0.23}{2} A_o \sqrt{h_o}$, where $\dot{m}_{O_2}$ is an empirical correlation for the mass flow rate of oxygen (kg/s), A_o and h_o are the area and height of the doorway opening (m^2), 0.23 is the mass fraction of oxygen in air. The parameter r in the table is the mass-based stoichiometric ratio of fuel to air required for complete combustion. The value of ϕ is useful in characterizing whether a given compartment fire is limited in size by its fuel supply or by its oxygen supply. The correlation for oxygen entrainment is valid for flashover conditions only, that is for values of $\phi > 1$.

In all of the experiments performed as part of this study, the fuel mass flow rate was either controlled (for the gaseous and liquid fuels) or measured (for the liquid and solid fuels), whereas the oxygen supply was naturally controlled by the size of the compartment doorway and the fire heat release rate. The range of values of ϕ varied as a function of fuel type, taking on values as small as $\phi = 0.1$ and as large as $\phi = 12.38$ as seen in Table 5.1. The value of ϕ was greater than 1.0 for almost all of the experimental conditions. This implies that conditions inside of the compartment were under-ventilated for the majority of the cases presented here. For each of the fuels listed in the table, except Nylon, flames were observed outside of the doorway, oxygen volume fractions were near zero and increased CO production was measured in the upper layer for the largest fires sizes. These are all strong indicators of under-ventilated burning. Pitts [52] proposed that a large fraction of the incoming air may be entrained into the out-flowing gases and never reach the reaction zone which indicates that the application of this relationship alone for scaling inadequate. A recent paper comparing results from the reduced scale enclosure [5] and experiments ISONG3 and ISOHept4, full door cases, also showed that this scaling factor does not always work [53]. According to the simple $Ah^{0.5}$ scaling relationship, the intermediate species observed at 250 kW in the reduced-scale enclosure, should be present at 1600 kW in the full-scale enclosure (FSE), however they were not present until approximately 2700 kW.

The third entry in the table is the compartment height, H, normalized by D^*, an effective characteristic burner dimension. The parameter H/D* relates $\dot{Q}$ to its physical dimensions and indicates the relative importance of the fire plume to other features of the fire-driven flow, such as the ceiling jet or doorway flow. This value is related to Q_d^* and similarities can be seen in the expressions for Q_d^* and H/D^* in Table 5.1. In contrast to the relative magnitude of Q_d^*, where larger values indicate a larger fire, larger values of H/D^* indicate a smaller fire in that the room is much larger than the fire size. Q_d^* only relates the fire size to the burner size while H/D^* relates the fire size to the dimension of the room. For this reason, the H/D^* parameter should give a better indication of the relationship of the fire size to the room size and be more effective for scaling predictions. Considering the case taken above for $Ah^{0.5}$ scaling [53], if the H/D^* scaling is used for the same cases, RSE: Q = 250 kW, H = 0.98 m; FSE: Q = 2700 kW, H = 2.4 m, and the RSE and FSE return values of H/D^* = 1.82 and H/D^* = 1.72, respectively. This indicates H/D^* is a better scaling relationship for these specific two tests than $Ah^{0.5}$ scaling. The range of values of H/D^* varied as a function of fuel type, taking on values as small as 2.17 and as large as 12.23 for conditions examined in this study, as seen in Table 5.1. For the reduced scale enclosure H/D^* had values in the range of 1.4 to 5.2.

Table 5.2 presents a comparison of scaling factors for three different heptanes fires including one fire from the RSE [5] and two heptane fires with different doorway openings in the FSE. Each case is compared at the heat release rate, Q, at which CO began to be detected at the front location in the room, indicating that the room was becoming under-ventilated and thus that there should be applicable under-ventilated scaling relationship coming into play. Previously a study that only considered the $Ah^{0.5}$ scaling was considered, and it was observed that for the full door FSE case the scaling relationship did not hold [53]. However if the 1/4 doorway (20 cm) case presented in Table 5.2 (ISOHept23) is considered with $Ah^{0.5}$ scaling the relationship is more applicable with the RSE and FSE where $Ah^{0.5}$ = 0.35 and $Ah^{0.5}$ = 0.57, respectively, predicting a heat release rate of 407 kW for the FSE, only a 40% error rather than the 60% error observed for the full door FSE case. The $Ah^{0.5}$ scaling works even better between the two FSE tests having values of $Ah^{0.5}$ = 2.26 for the full door case and $Ah^{0.5}$ = 0.57 for the quarter door case, resulting in only a 20% error in the predicted heat release rate between the two cases. Considering the other scaling factors the full doorway FSE case (ISOHept4) relates much better to the RSE case when H/D^* scaling is considered, the RSE and FSE having H/D^* = 1.82 and H/D^* = 1.72, respectively. However, when considering this scaling factor, the 1/4 doorway case no longer shows a scaling relationship to the RSE case, having H/D^* = 3.27. This indicates that the scaling factor that is used must be chosen judiciously in order to find accurate scaling of experiments. It also shows that it may be necessary to utilize multiple scaling factors to convert between one type of geometry and another, e.g. the full door RSE and the ¼ door FSE.

Table 5.2: Scaling comparison between the reduced scale enclosure (RSE) and the full scale enclosure (FSE) for heptane fires. The heat release rate values (Q) are taken as the heat release measured when CO began to be measured in the room.

Enclosure	Test ID	Q (kW)	H (m)	Doorway (h × w)	Q_d^*	$Ah^{0.5}$	ϕ	H/D^*
RSE	#15	250	0.89	0.81 × 0.48	2.11	0.35	0.49	1.81
FSE	ISOHept4	2700	2.4	2.0 × 0.8	9.82	2.26	0.83	1.71
FSE	ISOHept23	540	2.4	2.0 × 0.1	7.04	0.28	1.32	3.27

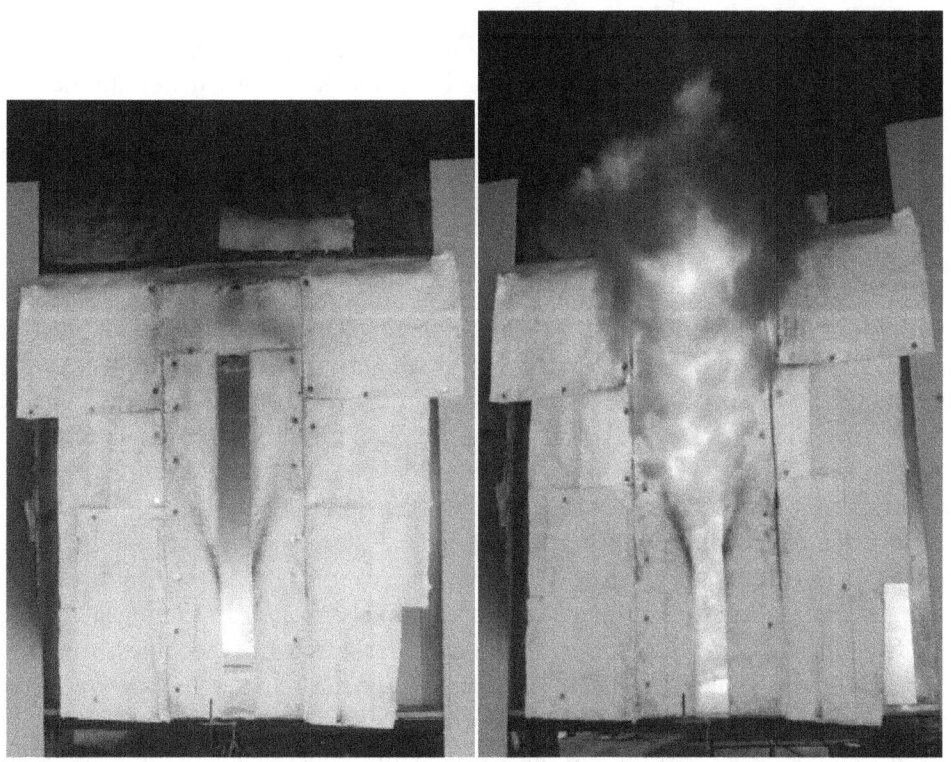

Figure 5.1: Example normalized scaling quantities with Q_d^* = 0.17, ϕ = 0.1, and H/D^* = 12.23, on the left (ISONylon10) and a Q_d^* = 12.74, ϕ = 7.53, and H/D^* = 2.17, on the right (ISOPropD14).

6 DISCUSSION

6.1 Behavior of different fuels

A variety of different fuels in three different phases were investigated here. Each fuel and each phase had its own unique challenges and behavior. This section endeavors to summarize some of the differences between individual fuels as well as between different fuel phases.

6.1.1 Liquid Fuels

In this investigation three liquid fuels were used. Heptane was used for many cases because it is a relatively well understood fuel that has been investigated extensively at NIST and by others. Heptane produces a moderate level of sooting and has a heat of combustion similar to gasoline. Iso-propanol was used because it is a different fuel, containing oxygen, and thus is expected to produce different results from heptane. Iso-propanol also produces a moderate level of soot but has a lower heat of combustion than heptane. Toluene was used as the third liquid fuel. Toluene generally produces copious amounts of soot and has a heat of combustion between that of heptane and iso-propanol.

ISOHept9, ISOProp15, and ISOToluene20 were three tests conducted under nearly identical conditions for the three fuels. For each case the 20 cm (1/4 width) door was used with a 0.5 m^2 pan burner situate in the center of the room at position 1, cf. Figure 2.5. For the heptane and iso-propanol tests 30 L of fuel was used and 20 L of fuel was used for the toluene test. In each case the fuel was ignited and allowed to burn freely until all the fuel was consumed. Figure 6.1, and Figure 6.3--Figure 6.6 present the heat release rate, temperatures, O_2, CO, and CO_2 time profiles respectively for the three liquid fuels considered here. Figure 6.1 presents the heat release rates of the three fuels as a function of time. The first observation that can be made is that the peak and average hat release rates of each fuel are qualitatively related to their heat of combustion. The heats of combustion, measured and ideal heat release rates of heptanes, toluene, and iso-propanol are listed in Table 6.1. There is a nearly linear relationship between the heat of combustion and the heat release rate for each of these fuels in the free burn. Figure 6.2 shows how the measured and ideal heat release rates vary with the net heat of combustion. Besides the nearly linear relationship it is also apparent from this plot that the combustion efficiencies of the different fuels may play a role here as well. However, despite heptanes having the largest heat release rate and heat of combustion and toluene having the lowest combustion efficiency the temperatures in the toluene fire were larger than those observed in either the heptane or iso-propanol fires, cf. Figure 6.3. The significantly higher temperature of toluene may be related to the chemistry happening within the enclosure. Considering Figure 6.4--Figure 6.6 the toluene fire has higher oxygen concentrations than either heptanes or iso-propanol, cf. Figure 6.4, along with less CO, cf. Figure 6.5, and more CO_2, cf. Figure 6.6. This indicates that the fire is not becoming as under-ventilated with toluene as it is with either heptane or iso-propanol. The increased temperature of the toluene fire can then be explained as the global equivalence ratio of the room being closer to stoichiometric conditions where fire temperatures are higher. The decrease of flame temperatures as the fuel/air mixture moves far beyond stoichiometric is a well understood fundamental phenomenon due to the increased quantity of unburnt fuel that absorbs energy from the reaction. A similar observation was made for the RSE experiments [5] and will be noted here in section 6.4 when the time varying heat release rate is discussed. Therefore, the reduced combustion efficiency indicates that much less than the ideal amount of fuel is being consumed and there is sufficient extra oxygen to achieve more complete combustion (more CO_2

and less CO) of the fuel that does burn. However, Figure 6.6 indicates that the toluene fire also has a very low volume fraction of hydrocarbons being sampled from the fire, however there is significant loss of fuel that is neither burned nor turned to detectable hydrocarbons. It is theorized that the majority of the unburnt toluene was converted to soot and was caught on the soot filters which were used to condition the gas sample flow.

Figure 6.8 and Figure 6.9 present a comparison of the heat flux being measured from three different liquid fuels: heptanes, isopropanol, and toluene. For each of the three heat flux gauges located in the ceiling the highest heat flux was realized from heptanes, cf. Figure 6.8. At the same time, for each of the two heat flux gauges located on the floor inside the room, toluene produced the largest heat flux, cf. Figure 6.9. One reason for this may be the chemical composition of the combustion products in the room. Figure 6.6 presents the comparison of the CO_2 volume fraction measured for each of these three fuels and shows that the CO_2 volume fraction for toluene is significantly higher than that for either heptanes or isopropanol. Therefore, one explanation for toluene having a higher heat flux to the floor than the other fuels is that its product gases are more heavily laden with CO_2. CO_2, being heavier than most of the other expected product gas constituents, is more likely to move the product gasses closer to the floor and increase the heat flux there.

Table 6.1: Heat of combustion and measured and ideal heat release rates for liquid fuels.

	Net Δh (MJ/kg) [3]	Measured HRR (kW)	Uncertainty of HRR	Ideal HRR (kW)	Uncertainty of Ideal HRR
Heptane	44.56	1460	105	1816	134
Toluene	40.52	1209	59.1	1567	155
Iso-Propanol	30.45	1111	75.9	1268	47

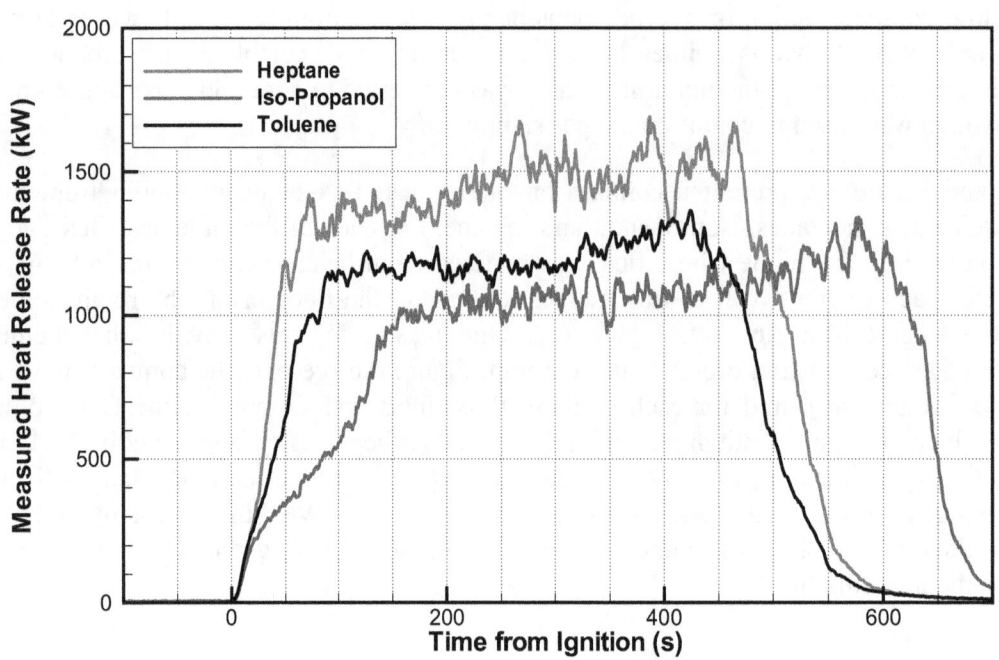

Figure 6.1: Comparison of measured heat release rate for three different liquid fuels, Heptane, Iso-Propanol, and Toluene.

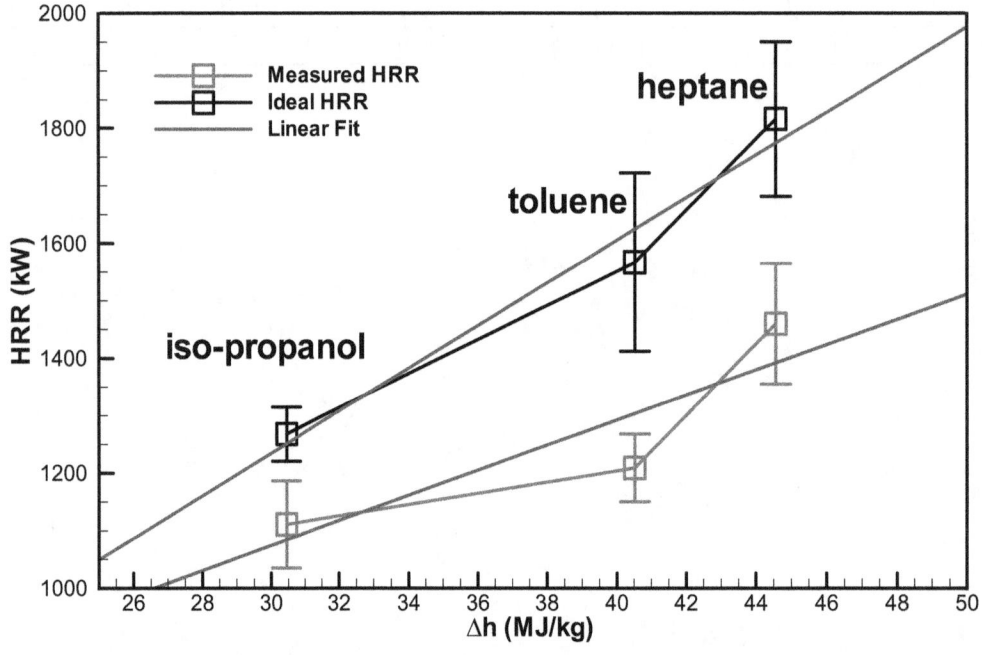

Figure 6.2: Plot of heat of combustion of each fuel verses measured and ideal heat release rate for three liquid fuels.

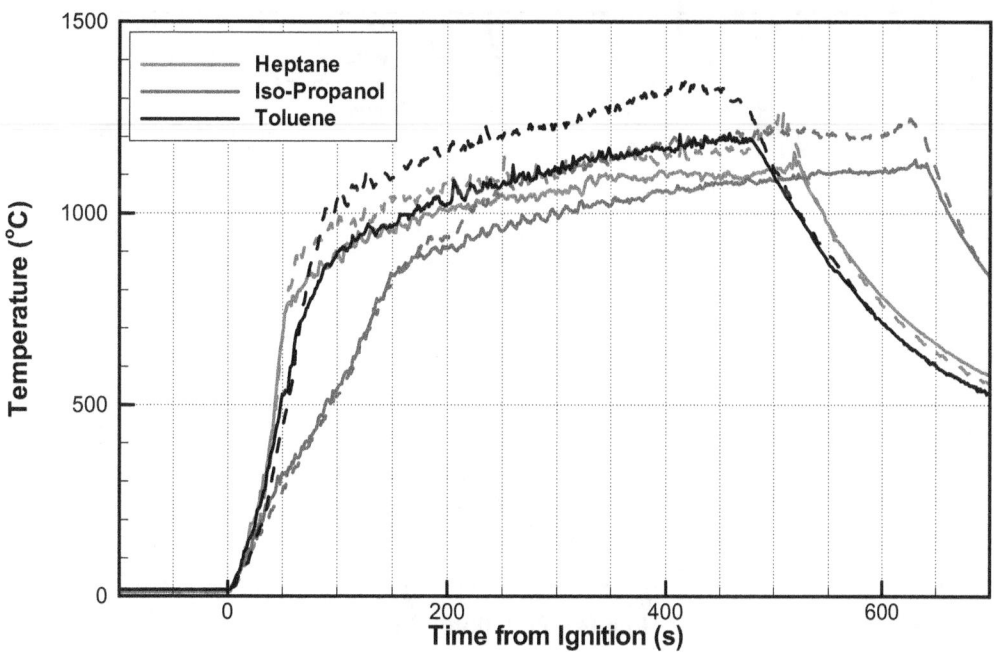

Figure 6.3: Comparison of measured front (solid lines) and rear (dashed lines) temperatures for three different liquid fuels, Heptane, Iso-Propanol, and Toluene.

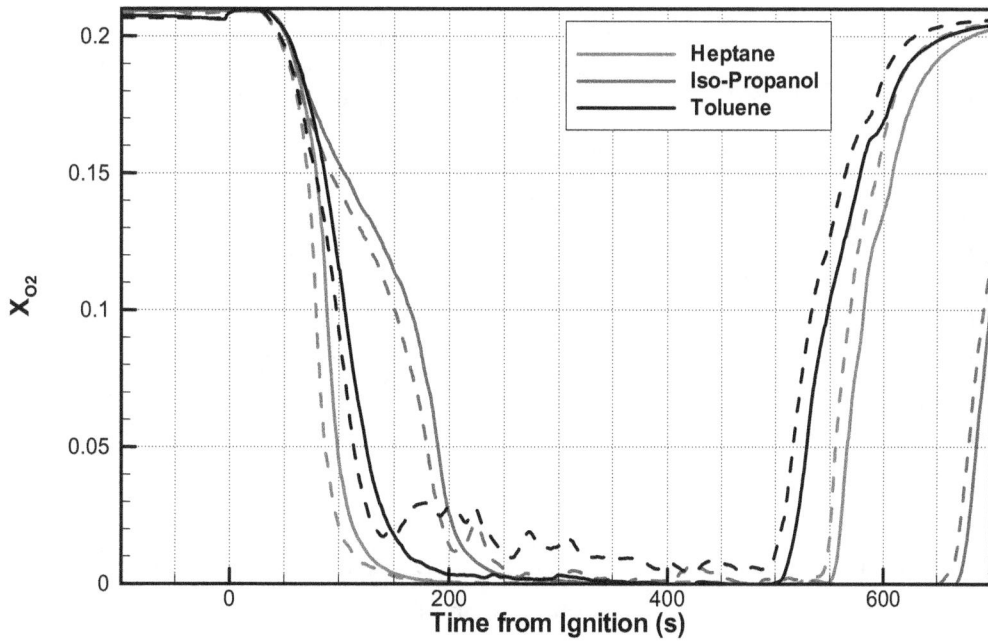

Figure 6.4: Comparison of measured front (solid lines) and rear (dashed lines) oxygen volume fraction as a function of time for three different liquid fuels, Heptane, Iso-Propanol, and Toluene.

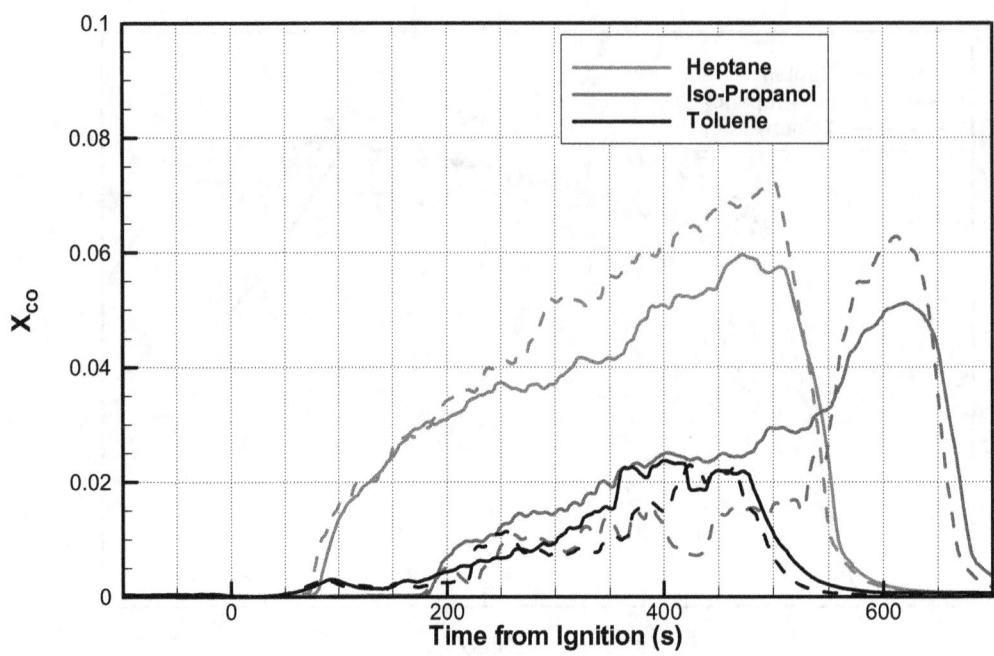

Figure 6.5: Comparison of measured front (solid lines) and rear (dashed lines) CO volume fraction as a function of time for three different liquid fuels, Heptane, Iso-Propanol, and Toluene.

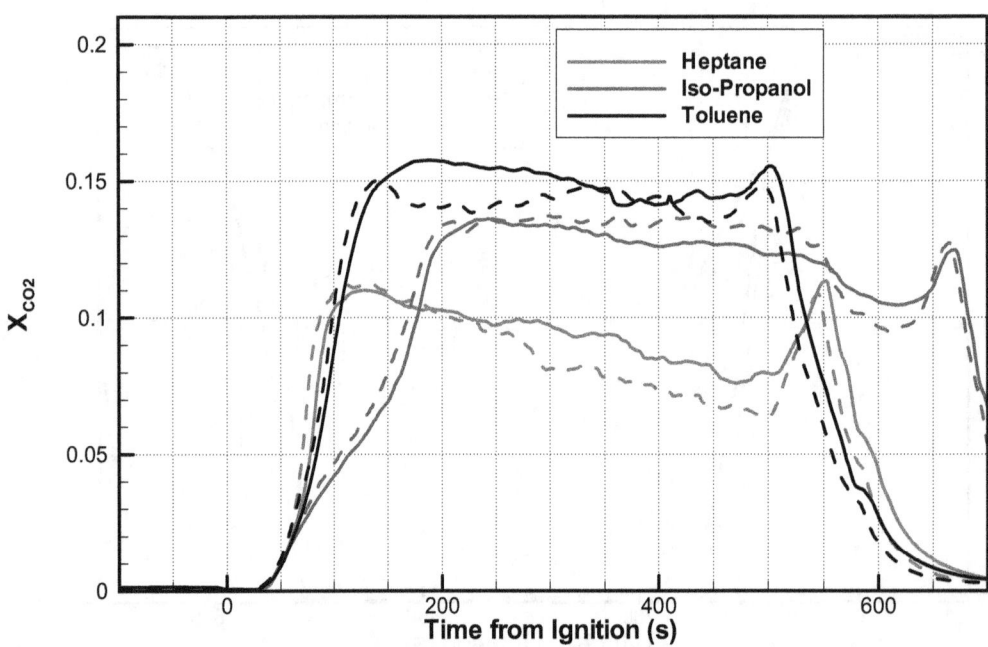

Figure 6.6: Comparison of measured front (solid lines) and rear (dashed lines) CO_2 volume fraction as a function of time for three different liquid fuels, Heptane, Iso-Propanol, and Toluene.

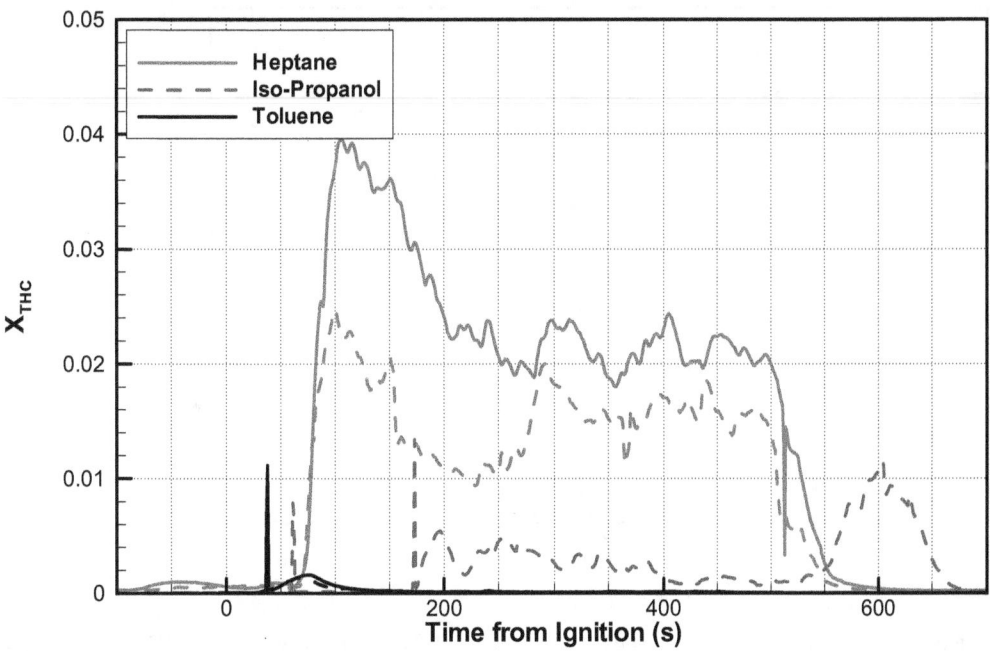

Figure 6.7: Comparison of measured front (solid lines) and rear (dashed lines) total hydrocarbon (THC, on a CH_4 basis) volume fraction as a function of time for three different liquid fuels, Heptane, Iso-Propanol, and Toluene. (Note: Front Iso-propanol results are not shown as the analyzer failed during this test.

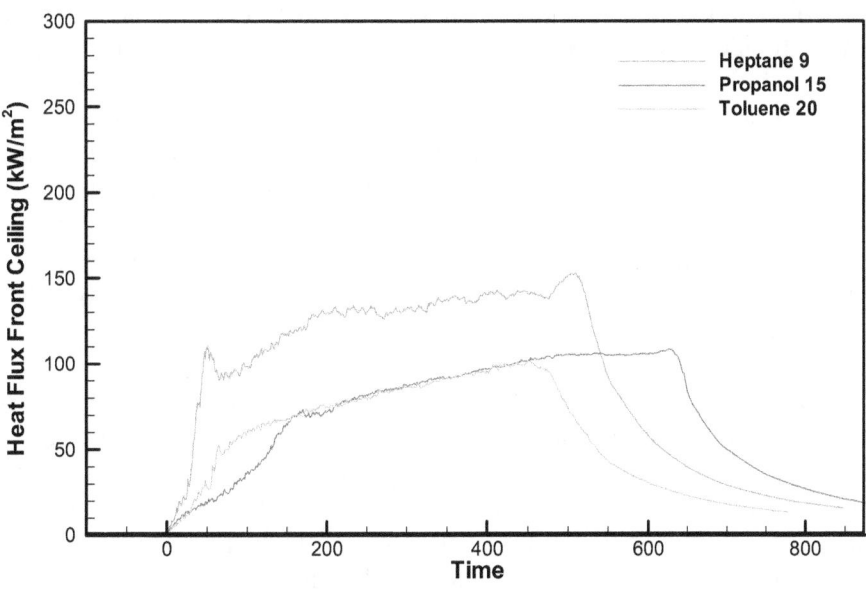

Figure 6.8: Comparison of the measured radiative heat flux at the front ceiling location, channel ID:HFFCE, for heptanes, toluene, and isopropanol.

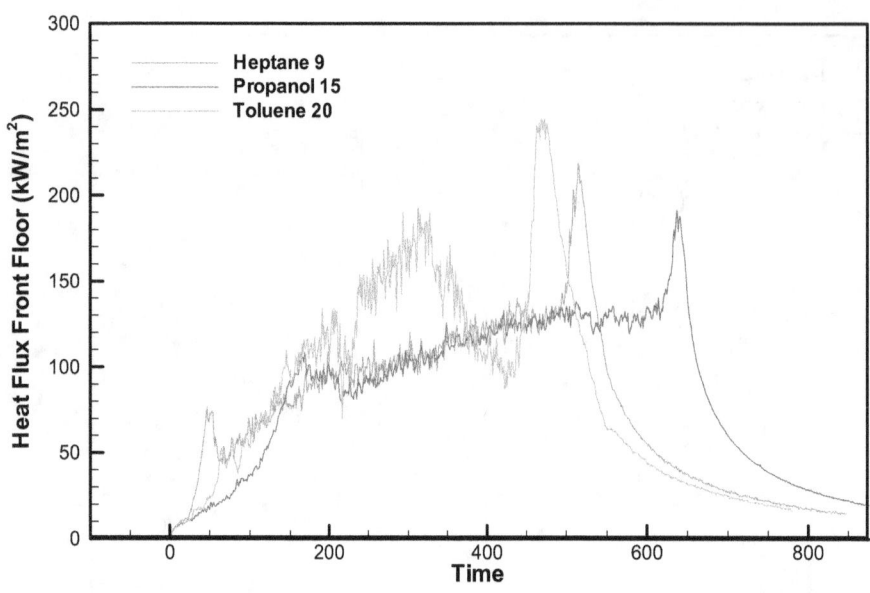

Figure 6.9: Comparison of the measured radiative heat flux at the front floor location, channel ID:HFFFL, for heptanes, toluene, and isopropanol.

6.1.2 Solid Fuels

In the configuration described in this report, creating under-ventilated fires with solid fuels while still being able to measure ideal heat release rate as the mass loss rate of the fuels was problematic. For this reason only two under-ventilated solid fuel fires were successfully created. ISOStyrene17 and ISOPP18 using 30 kg polystyrene and 20 kg polypropylene, respectively, are those two cases. The cases are similar in that they each used the same door opening (20 cm) and utilized the same surface area of solid fuel, and similar masses of fuel. However, the two tests differ in that the polystyrene fire utilized a single 1 m^2 burner positioned in the center of the room at position 1, cf. Figure 2.5, and the polystyrene fire utilized two 0.5 m^2 pans at positions 1 and 2. Despite these differences in configuration there are some interesting comparisons that can be made between the two solid fuels.

Figure 6.10 presents the measured heat release rate for the polystyrene and polypropylene fuels. The polystyrene fire heat release rate initially increases very quickly compared to the polypropylene fire. Later the polystyrene heat release rate peaks earlier than the polypropylene heat release rate. The earlier maximum in heat release rate for polystyrene can be attributed to the pan configuration. All of the polystyrene fuel is contiguous while the polypropylene fuel was in two separate locations that initially did not interact much. The initially faster rise in heat release rate for polystyrene however is more likely to occur because of a smaller heat capacity of polystyrene. Additionally, it is observed that the maximum heat release rate of the polypropylene fire is higher than that of the polystyrene fire. This is likely because the net heat release rate for polypropylene is larger than polystyrene.

Figure 6.11 compares the temperatures of polypropylene and polystyrene as a function of time. Analogous to the heat release rate the temperature of the polystyrene fire initially increases faster than that of the polypropylene fire. The maximum temperature of the polystyrene fire occurs earlier than the polypropylene fire as well, however the maximum temperature of the polystyrene fire is identical to that of the polypropylene fire within the experimental uncertainty. This can again be related to the heat capacity of the two materials. The polypropylene has a larger heat capacity than the polystyrene which negatively influences the temperature of the fire. So that even though polypropylene has a higher heat release rate than polystyrene the temperatures are nearly equal.

Figure 6.12 -- Figure 6.15 illustrate the chemistry of the two fires. These plots illustrate the expected behavior. Specifically, O_2 decreases while CO_2 and CO increase earlier for polystyrene. Then the polypropylene fire has a lower O_2 and CO_2 volume fractions with a larger CO and THC volume fractions than the polystyrene flame. This corresponds to the larger heat release rate shown in Figure 6.10. When the heat release rate is higher more oxygen is necessary for complete combustion, however the same ventilation is provided for both the polystyrene and polypropylene fires. Therefore, the polystyrene fire has more oxygen than the polypropylene fire. Less oxygen for the polypropylene fire results in lower O_2 volume fraction, lower CO_2 volume fraction, and higher CO and THC fractions because there is not enough oxygen to convert CO to CO_2 or even consume the extra fuel.

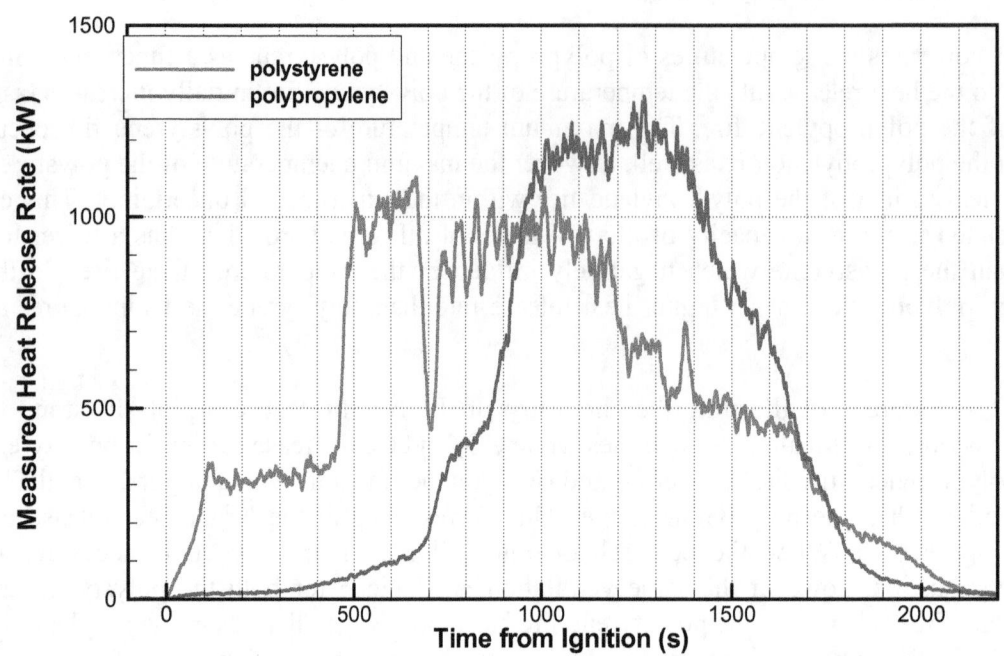

Figure 6.10: Comparison of the measured heat release rate for polystyrene and poly propylene solid fuels.

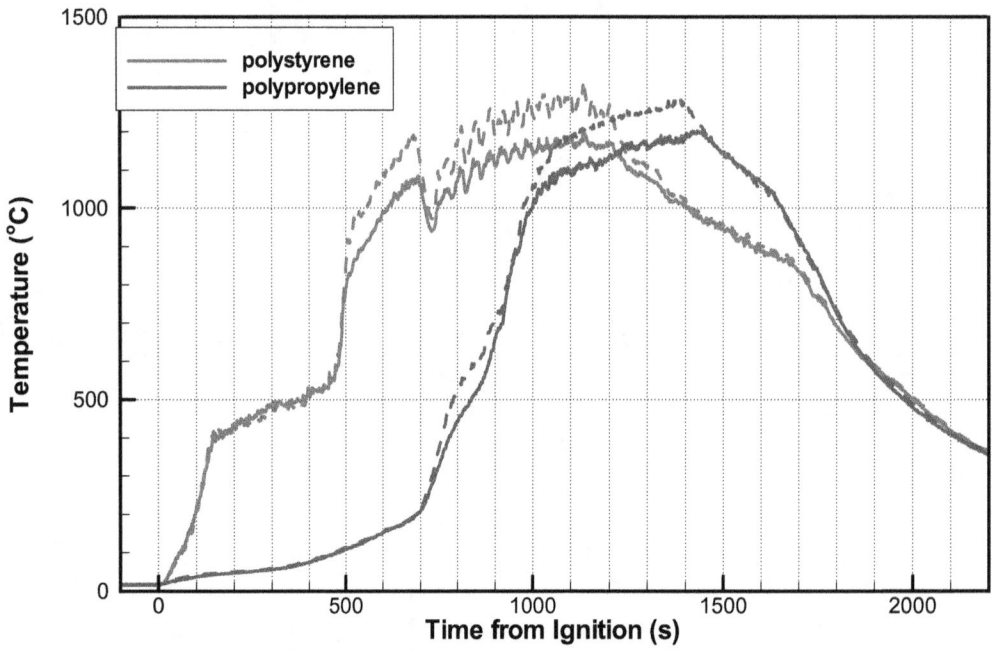

Figure 6.11: Comparison of measured front (solid lines) and rear (dashed lines) temperature as a function of time for two different solid fuels, polystyrene and polypropylene.

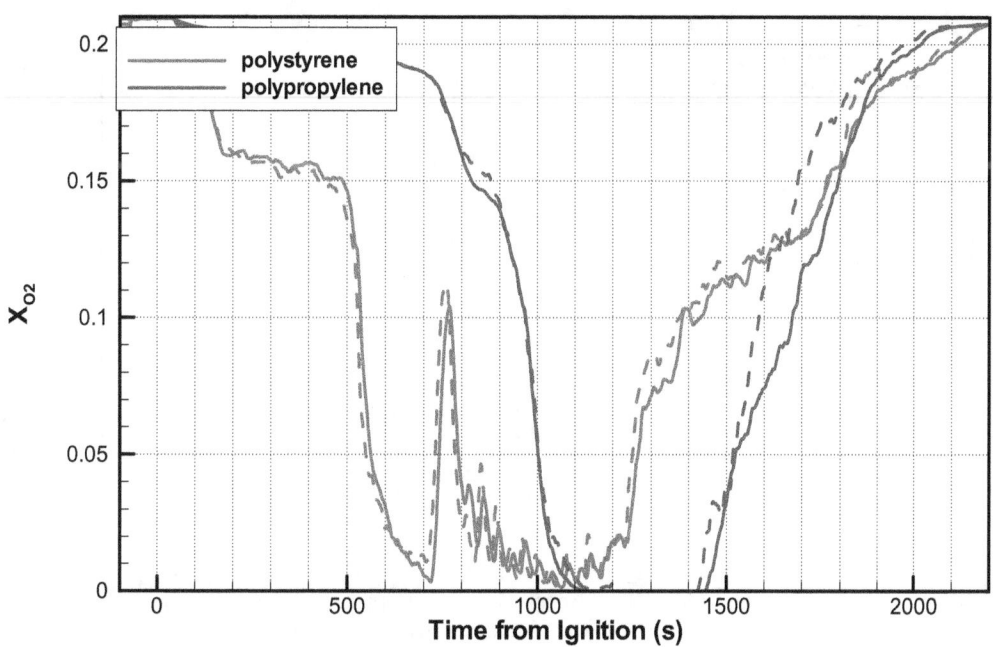

Figure 6.12: Comparison of measured front (solid lines) and rear (dashed lines) oxygen volume fraction as a function of time for two different solid fuels, polystyrene and polypropylene.

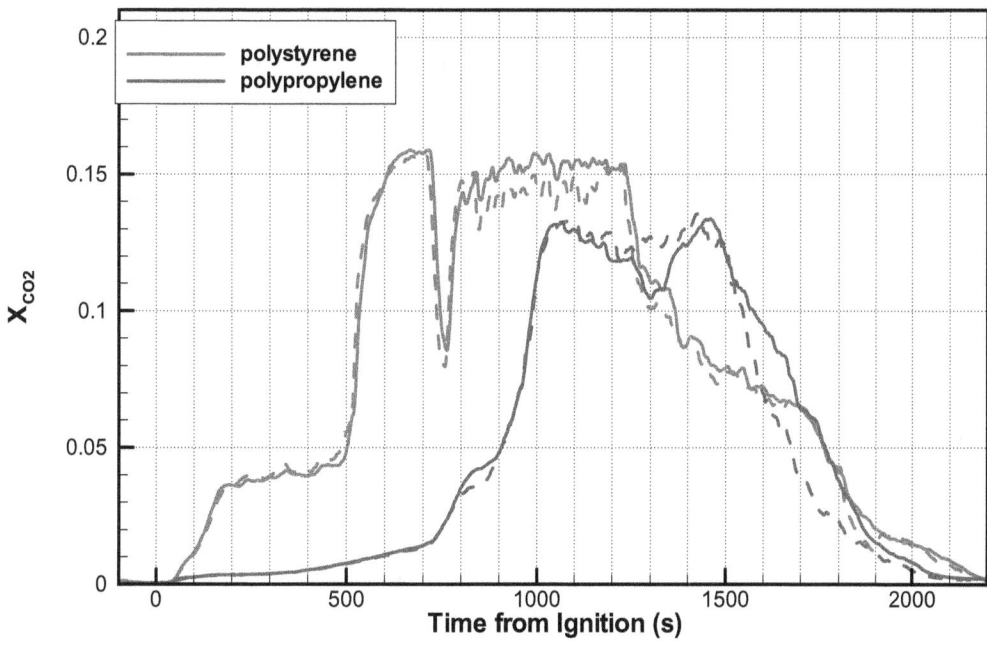

Figure 6.13: Comparison of measured front (solid lines) and rear (dashed lines) CO_2 volume fraction as a function of time for two different solid fuels, polystyrene and polypropylene.

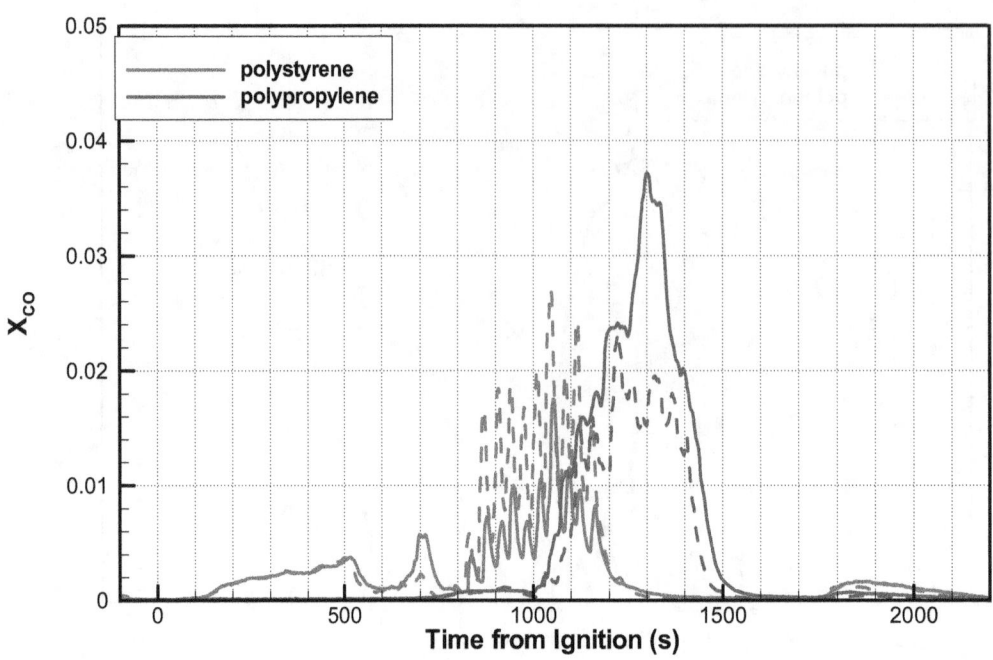

Figure 6.14: Comparison of measured front (solid lines) and rear (dashed lines) CO volume fraction as a function of time for two different solid fuels, polystyrene and polypropylene.

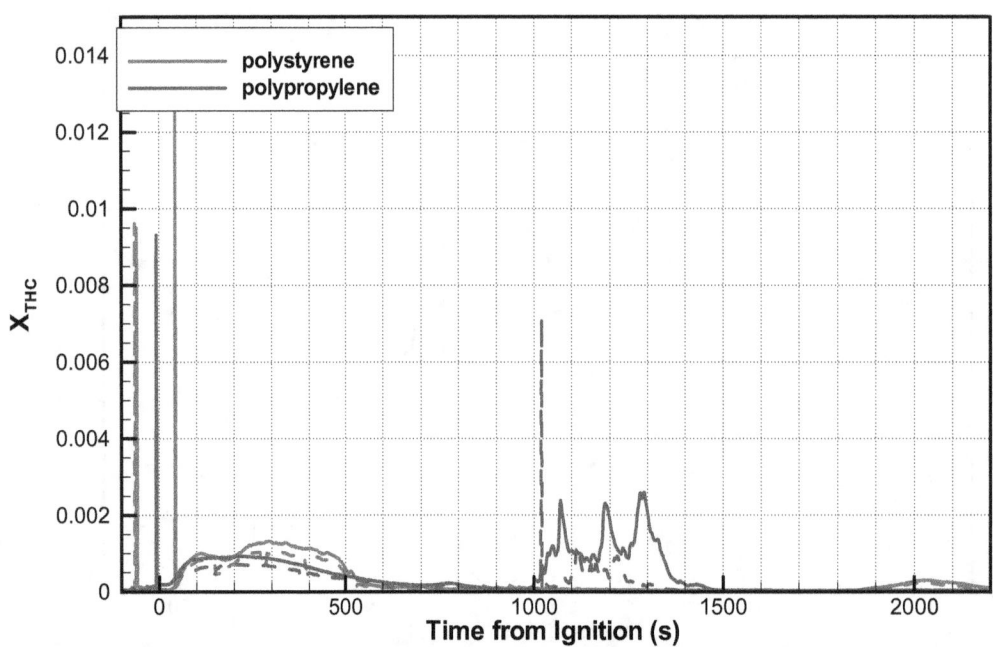

Figure 6.15: Comparison of measured front (solid lines) and rear (dashed lines) total hydrocarbon volume fraction (on a CH_4 basis) as a function of time for two different solid fuels, polystyrene and polypropylene.

6.1.3 Comparison

There are several significant differences between the combustion of solid and liquid fuels that are worth noting here. First the liquid fuels always start burning faster. This is because the liquid fuels have a much smaller effective heat capacity than the solid fuels. The liquid fuels must only vaporize to burn, while the solid polymer fuels first melt then vaporize and burn. Overcoming this initial heat capacity takes more time for the solid fuels. Secondly, the radiative feedback to the fuel source is important. All of the liquid fuels became under-ventilated easily with a 0.5 m^2 pan while a total area of 1 m^2 was necessary to cause the solid fuels to consume all of the oxygen in the room. This again is because of the significantly larger heat needed to actually burn the solid fuels. In effect the solid fuels need to recycle some of their heat of combustion to the fuel surface to sustain combustion. Convection and conduction are insufficient in this case because the cool air is entering the room from the doorway and cooling the pan burner so that the additional heat needed to melt and vaporize the solid polymer fuels must come from radiative feedback from the high temperature upper layer of the room. A list of some general fuel properties is provided in Table 6.2.

Table 6.2: Fuel Properties (r is the stoichiometric ratio of fuel to air).

	Formula	ρ (g/cm^3)	C_p (J/g°C)	Δh_{net} (kJ/g)	r
Natrual Gas	~CH$_4$	7×10^{-4}-9×10^{-4}	2.23	50.03	0.058
Heptane	~C$_7$H$_{16}$	0.684	2.2	44.56	0.066
Toluene	C$_7$H$_8$	0.866	1.67	40.52	0.075
Iso-propanol	C$_3$H$_8$O	0.786	2.42	30.45	0.101
Polystyrene	C$_8$H$_8$	1.05	1.4	39.75	0.076
Polypropylene	C$_3$H$_6$	0.855	2.1	43.23	0.068
Nylon	C$_6$H$_{11}$NO	1.15	1.5	29.3	0.102

6.2 Effect of fuel distributions

In this report fuel distribution refers to having the fuel sources, or burners located at more than one position inside the room. Most often the fuel distributions that were studied here were a comparison of a single 0.5 m^2 burner located at position 1, cf. Figure 2.5, and two 0.25 m^2 burners located at positions 1 and 2, cf. Figure 2.5. Figure 6.16 -- Figure 6.22 compare several quantities from heptane fires in a single burner (ISOHept9) and in two, distributed, burners (ISOHept12) as described above.

Figure 6.16 compares the heat release rates of the one and two burner configurations. The average heat release rate from each configuration is nearly the same, which is to be expected since an identical quantity of fuel with identical surface area was used for each configuration. The differences between the two configuration is in their transient behavior. The one burner case heat release rate is initially smaller than the two burner case and gradually increases to a maximum near the end of the experiment. The two burner case initially peaks in heat release rate early and then proceeds to gradually decrease before dropping off rapidly after the fire burns out.

Figure 6.17 presents a comparison of the front and rear temperatures for the one and two burner cases. The front temperatures show a small difference (~50 °C) however the rear temperatures are nearly identical which indicates that the thermal environment of the two tests was nearly identical.

Figure 6.18 presents a comparison of the front and rear O_2 volume fractions for the one and two burner tests. Similarly to the temperature results presented in Figure 6.17, the one and two burner cases are nearly identical in their O_2 volume fraction profiles. The only difference in O_2 volume fraction of note is the small increase in the rear O_2 volume fraction prior to the fire going out for the two burner case. In this case the rear burner exhausted its fuel prior to the center burner. This may have allowed more oxygen to move and be present in the rear of the enclosure.

Figure 6.19 and Figure 6.20 present comparisons of the CO_2 and CO volume fraction profiles respectively. A more pronounced difference between the one and two burner cases can now be seen. More CO_2 and less CO are present for the two burner case. This could indicate that the two burner case is more well ventilated than the one burner case causing it to produce more CO_2 and less CO. However when Figure 6.21 is considered the total hydrocarbons in the front of the enclosure are much higher for the two burner case than the one burner case. These results seem to counter indicate each other, however the higher total hydrocarbons in the front of the enclosure may be the result of vaporized fuel (from the center burner) being pushed out of the enclosure so quickly that it does not have time to burn completely. Another indication of this is the heat flux measured directly over the center burner. Figure 6.22 presents a comparison of the front floor (between the center burner and the doorway) and the center ceiling (directly above the center burner) heat flux gauges for the one and two burner cases. The front floor heat flux is largely similar for the two cases. The center ceiling heat flux shows a dramatically different behavior. For the one burner case the center ceiling heat flux is much larger than the front floor heat flux while for the two burner case the center ceiling heat flux is much smaller than the front floor heat flux. This is in agreement with the previously presented theory of blowing vaporized fuel out of the enclosure before it can burn. As the combustion products from the rear burner move over the center burner they heat the fuel and vaporize it, however all of the oxygen has already been consumed. Therefore the heat present in the room above the center burner is much smaller due to its being absorbed by the fuel which is not burning. It is likely that the excess fuel burns once it exits the enclosure as evidenced by the nearly equal heat release rates from the two cases.

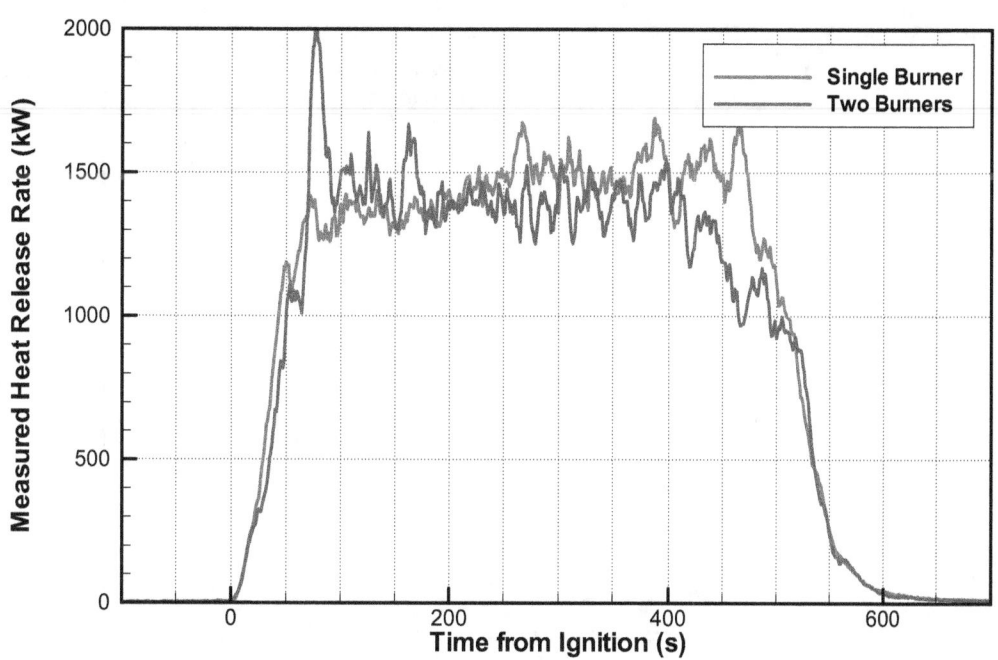

Figure 6.16: Comparison of heat release rate for heptane fires with a single burner and two distributed burners.

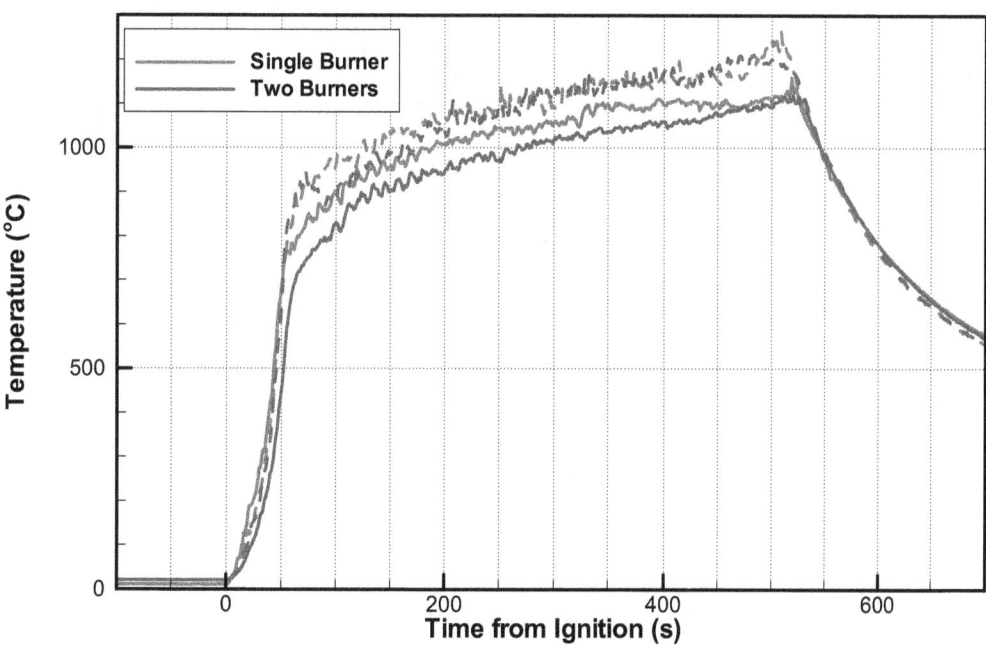

Figure 6.17: Comparison of front (solid line) and rear (dashed line) temperatures for heptane fires with a single burner and with two distributed burners.

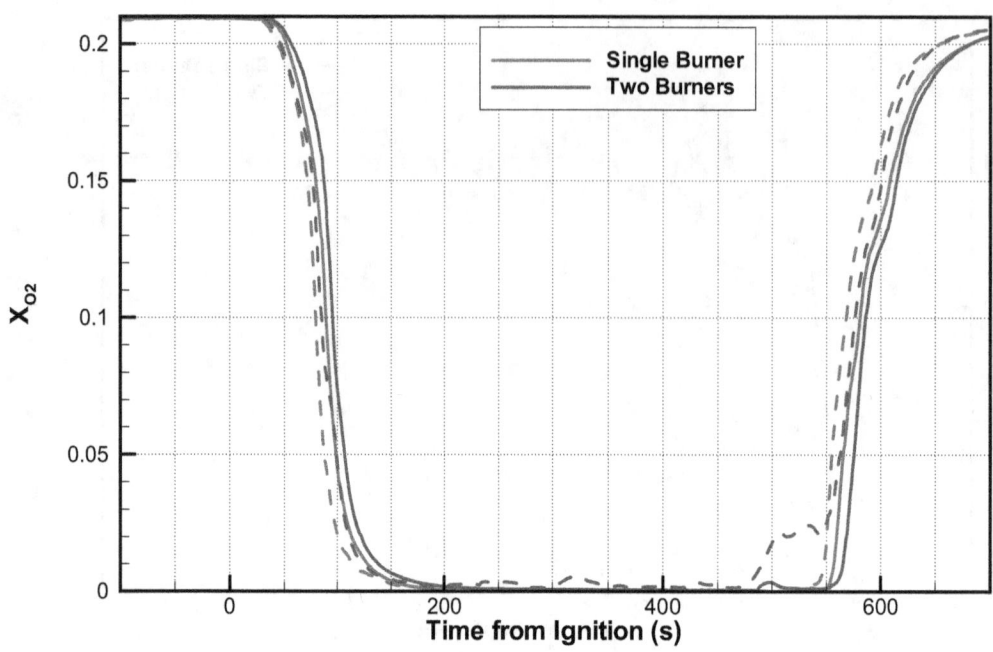

Figure 6.18: Comparison of front (solid line) and rear (dashed line) O_2 volume fraction for heptane fires with a single burner and with two distributed burners.

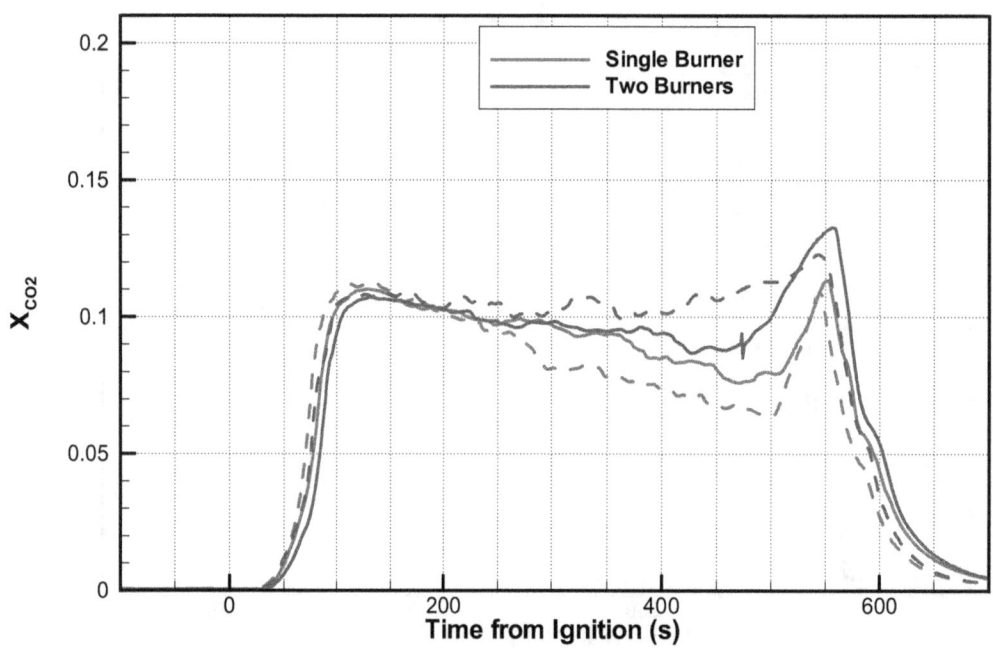

Figure 6.19: Comparison of front (solid line) and rear (dashed line) CO_2 volume fraction for heptane fires with a single burner and with two distributed burners.

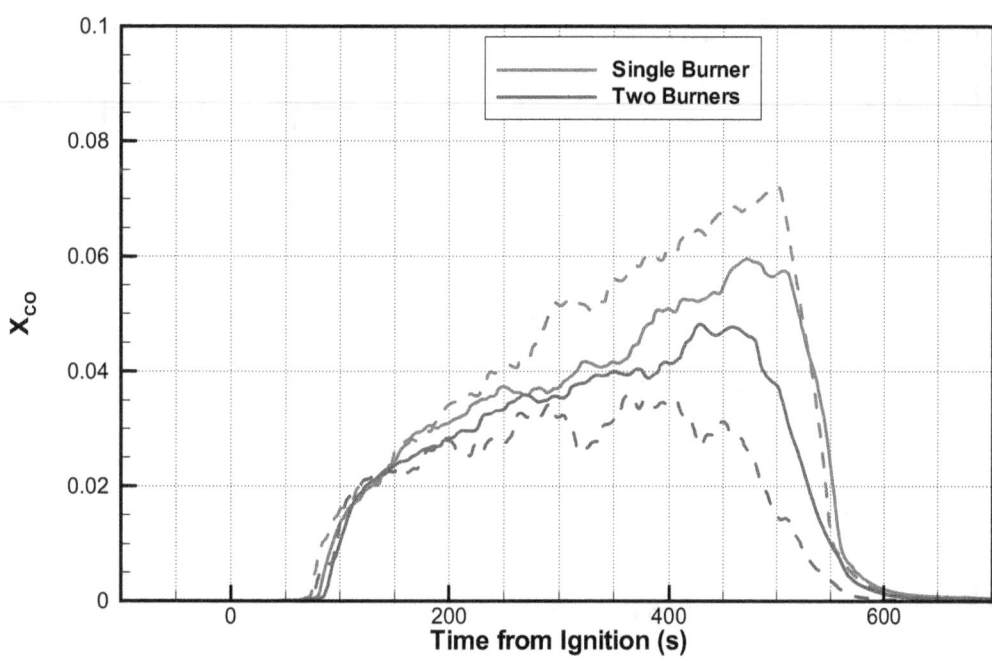

Figure 6.20: Comparison of front (solid line) and rear (dashed line) CO volume fractions for heptane fires with a single burner and with two distributed burners.

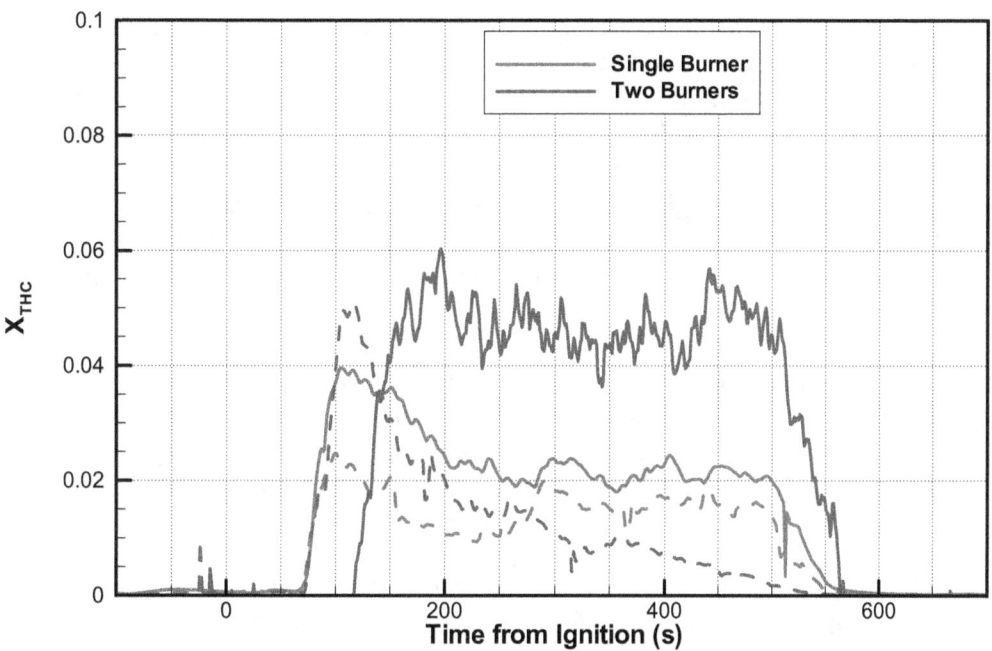

Figure 6.21: Comparison of front (solid line) and rear (dashed line) total hydrocarbon volume fraction for heptane fires with a single burner and with two distributed burners.

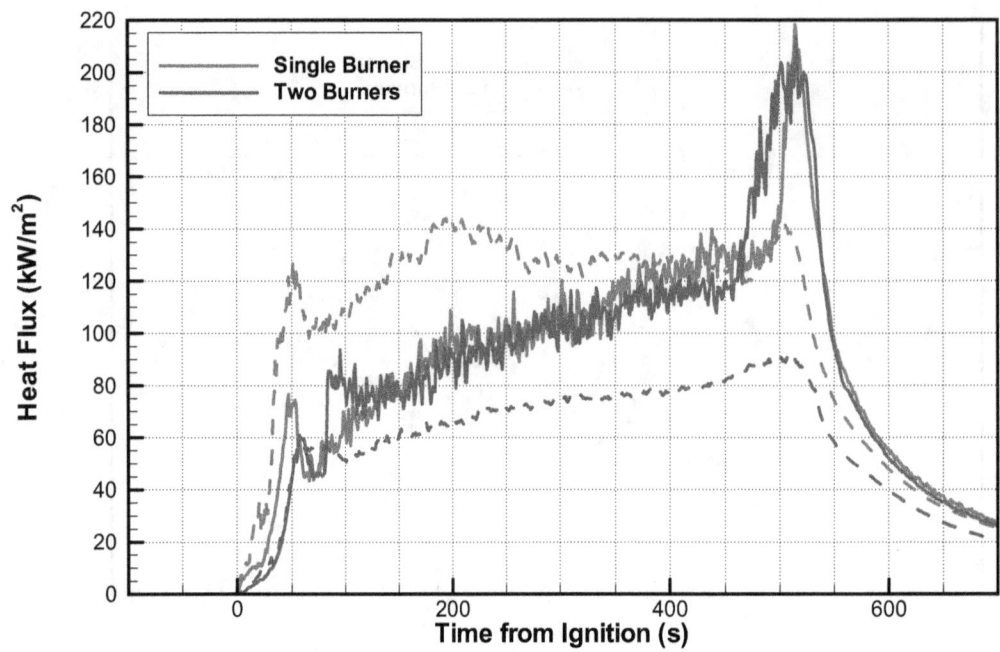

Figure 6.22: Comparison of front floor (solid line) and center ceiling (dashed line) heat fluxes for heptane fires with a single burner and with two distributed burners.

6.3 Ventilation effects

Figure 6.23 -- Figure 6.28 present comparisons of several key variables for three heptane fires with various door widths to vary the ventilation of the enclosure. ISOHept25, ISOHept22, and ISOHept24 represent heptanes fires where a spray burner was used; the door width was varied in fractions of the full scale (80 cm) ISO 9705 room. ISOHept25 utilized a 1/2 width (40 cm) doorway. ISOHept22 utilized a 1/4 width (20 cm) doorway, which was also used for the majority of the other experiments in this test series. ISOHept24 utilized a 1/8 width (10 cm) doorway. Each test used the full doorway height (2 m) dimension of the ISO 9705 enclosure. Each test stepped through a series of steady state heat release rates, the heat release rate of 1000 kW was common to each test and the pseudo-steady period when that heat release rate was measured is presented in Figure 6.23 -- Figure 6.28.

Figure 6.23 presents a comparison of the measured heat release rate for each of the three ventilation cases. The measured heat release rate was very similar for each of the cases over the duration of the pseudo-steady state region that is presented. This is by design and provides confidence for the ensuing discussion that the heat release rate is actually the same for each case and only the vent dimensions of the enclosure are changing.

Figure 6.24 presents a comparison of the temperatures in the three different ventilation conditions. A slight transient is evident for the 40 cm door case, however the temperatures in both the front and rear of the room are similar between the three ventilation cases. This may be explained by the fact that the heat release rate is being maintained for each case so that the total heat being produced in the room is similar and the same thermal environment should be present.

However, as will be seen later the chemical environment of the room was much different among the cases.

Figure 6.25 presents a comparison of the oxygen measurements at the front and rear of the room for each of the three cases. Approximately 10 % oxygen is present in the 40 cm door case while there was approximately 3 % oxygen in the rear of the 20 cm door and no measured oxygen for the other cases. This indicates that even though the same heat release rate was present and similar temperatures were observed the 40 cm case was not ventilation limited while the 20 cm and 10 cm cases were sequentially more ventilation limited with decreasing vent size.

Figure 6.26, Figure 6.27, and Figure 6.28 present the CO_2, CO, and total hydrocarbon (THC) volume fractions respectively for each ventilation condition. The 10 cm and 40 cm doorway cases have similar volume fractions of CO_2 while the 20 cm doorway has significantly more CO_2. This seems contradictory in comparison to the heat release rate and oxygen data where it might be expected to produce more CO_2 in the 10 cm doorway case as more O_2 is consumed. However, considering Figure 6.27 and Figure 6.28 it is evident that the carbon that would have been going to CO_2 production is instead being used to produce CO and is resulting in a larger volume fraction of THC because all of the oxygen has been used. Thus, the 10 cm doorway case can have less CO_2 than the 20 cm doorway case and the same CO_2 volume fraction as the 40 cm doorway case when there is significantly less oxygen present in the room.

This brief investigation into the effects of varying the ventilation of a free burning fire in a room provides insight into how the enclosure fire behaves. Reducing the vent size with the same size fire is equivalent to increasing the fire size in a room with a larger vent size. Both of these scenarios contribute to producing an under-ventilated condition within the enclosure. This idea is expanded upon in the next section (section 6.4) where an enclosure with a fixed vent size has the heat release rate increased linearly with time so that the evolution of a room from well ventilated to under-ventilated can be observed.

A correlation such as the ventilation scaling relationships discussed in section 5 may be used to correlate effective ventilation conditions. As an example consider the oxygen volume fractions presented in Figure 6.25. The O_2 volume fraction being measured in the rear of the 40 cm doorway case is approximately twice the value measured in the rear of the 20 cm doorway case. The ventilation scaling relationship says that the ventilation of the room is proportional to the factor $Ah^{0.5}$. Since this investigation considers only varying the width of the doorway the factor of proportionality is simply $wh^{1.5}$ or the width multiplied by a constant, for constant doorway height. In this context, reducing the width of the vent (doorway) by a factor of 2 results in a reduction in the ventilation of the room by 1/2 which is reflected in the 1/2 factor difference in the O_2 volume fraction between the 40 cm doorway case and the 20 cm doorway case. Similarly, consider Figure 6.27 where the CO volume fractions are presented. The CO volume fraction scales with the width of the vent. The 10 cm doorway has twice the CO volume fraction of the 20 cm doorway case, proportional the variation in ventilation, proportional here only to doorway width.

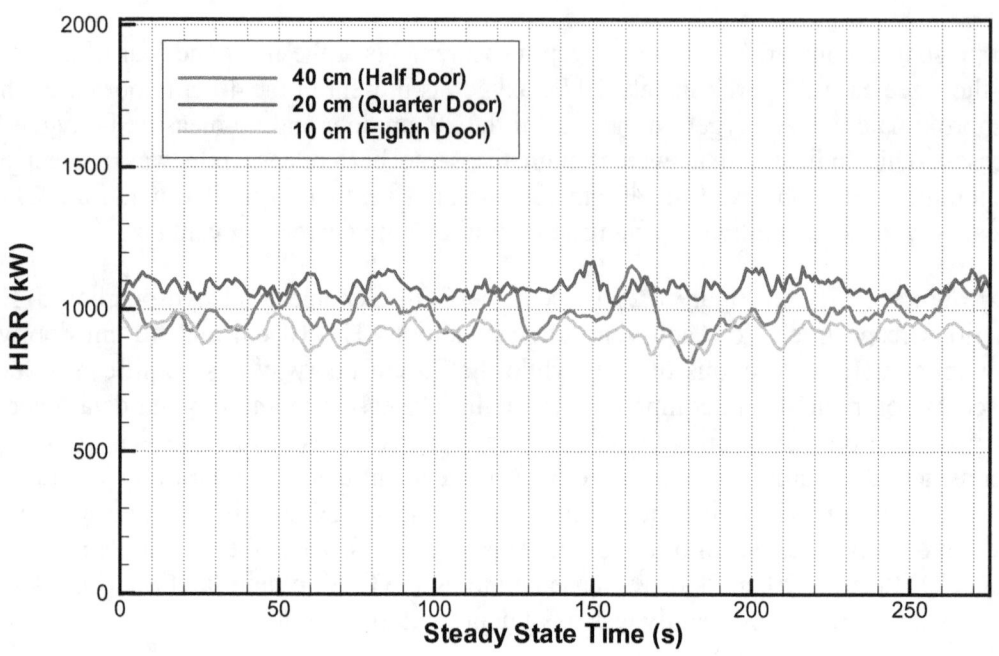

Figure 6.23: Comparison of heat release rates for three different ventilation conditions of a heptane spray burner fire operating at a nearly steady state of 1000 kW.

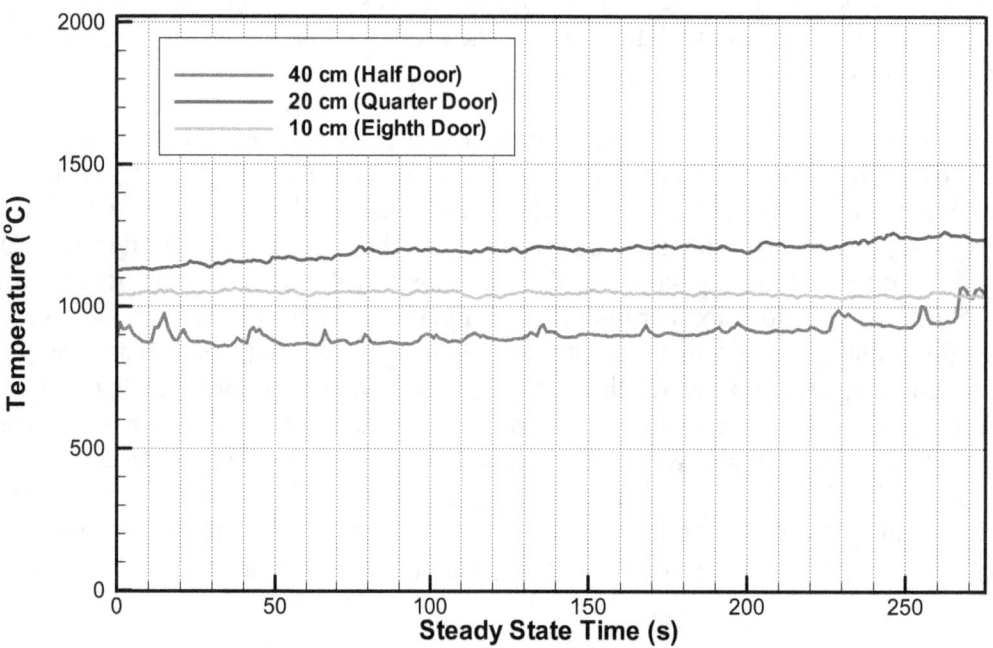

Figure 6.24: Comparison of front (solid likes) and rear (dashed lines) temperatures for three different ventilation conditions of a heptane spray burner fire operating at a nearly steady state of 1000 kW.

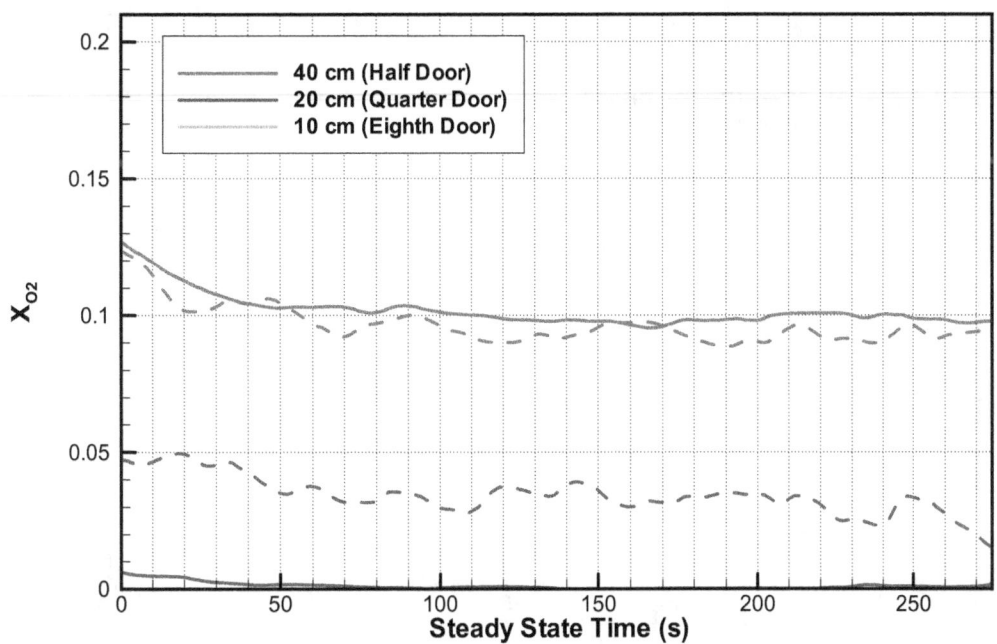

Figure 6.25: Comparison of front (solid likes) and rear (dashed lines) O_2 volume fraction for three different ventilation conditions of a heptane spray burner fire operating at a nearly steady state of 1000 kW.

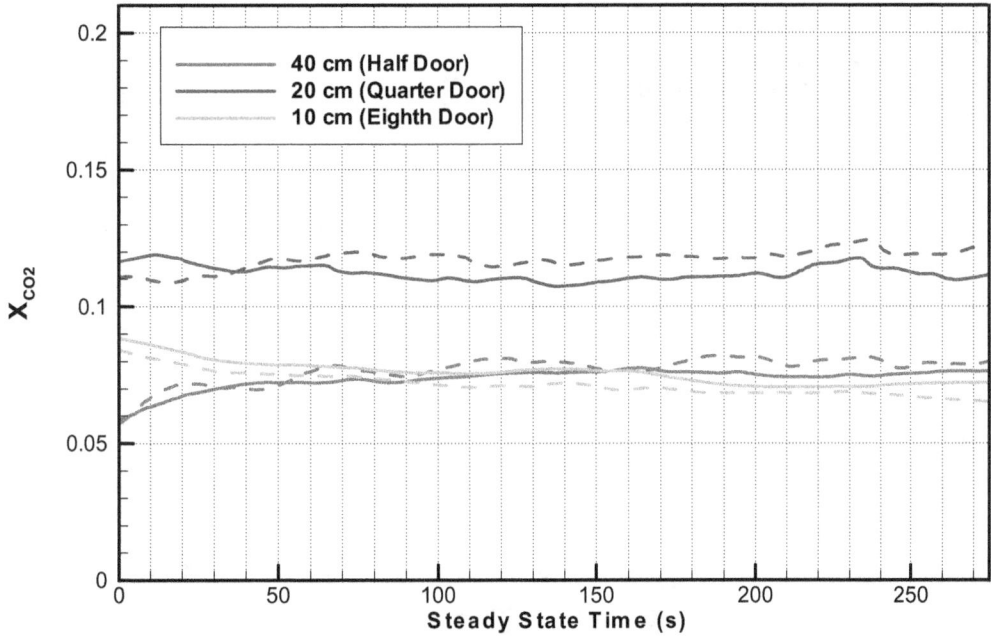

Figure 6.26: Comparison of front (solid likes) and rear (dashed lines) CO_2 volume fraction for three different ventilation conditions of a heptane spray burner fire operating at a nearly steady state of 1000 kW.

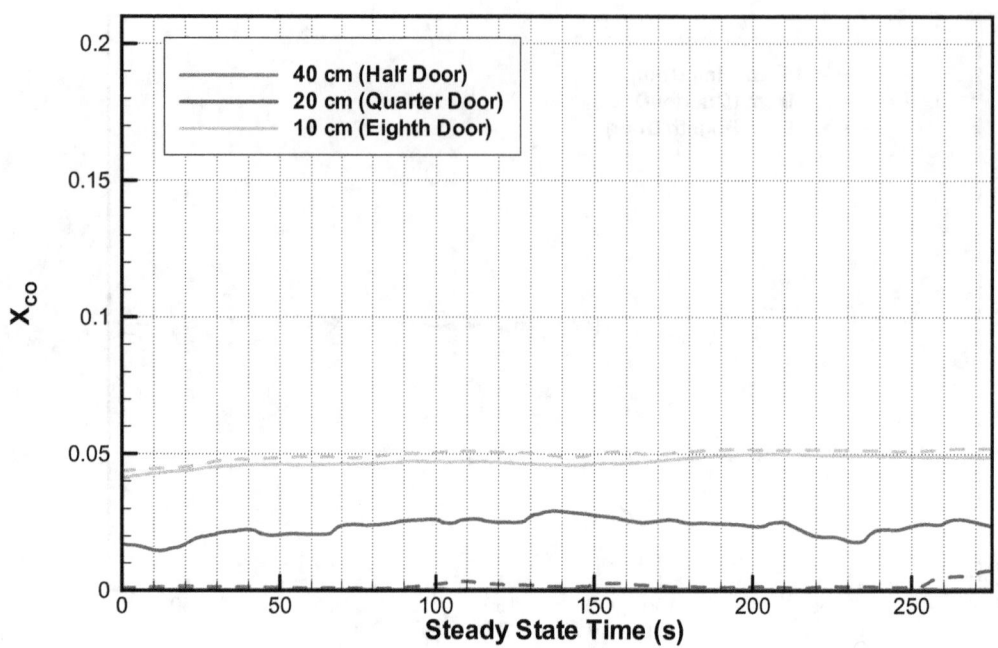

Figure 6.27: Comparison of front (solid likes) and rear (dashed lines) CO volume fraction for three different ventilation conditions of a heptane spray burner fire operating at a nearly steady state of 1000 kW.

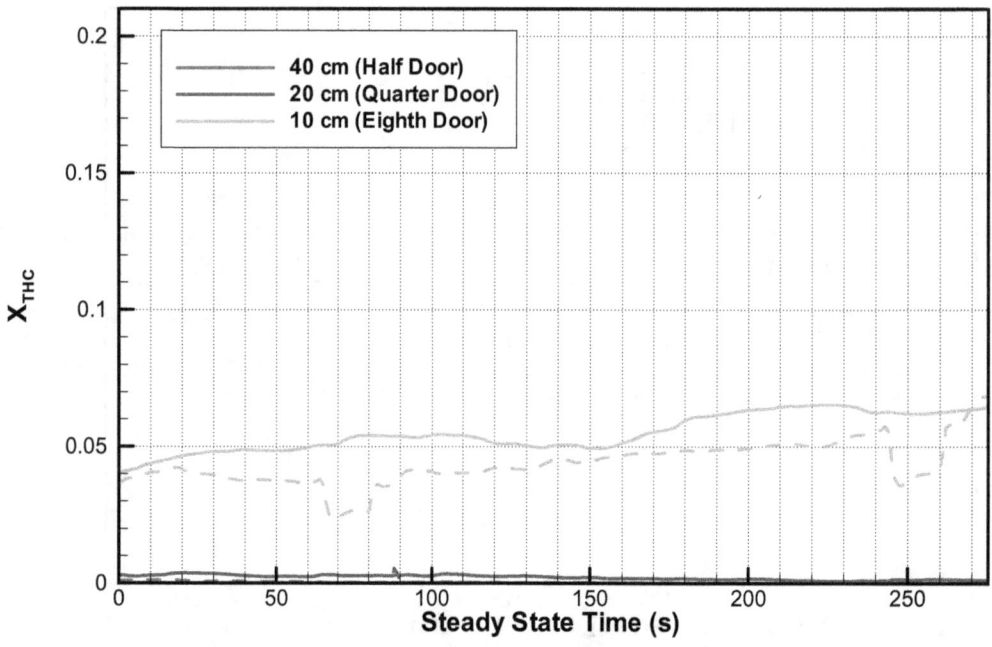

Figure 6.28: Comparison of front (solid likes) and rear (dashed lines) total hydrocarbon volume fraction for three different ventilation conditions of a heptane spray burner fire operating at a nearly steady state of 1000 kW.

6.4 Heat release rate ramp

Figure 6.29 and Figure 6.30 present the heat release rate, temperature, and species volume fractions for test ISOHept27 where heat release rate was ramped linearly with time with a 10 cm (1/8 width) door. This test is illustrative of how the environment inside the room is changing as a function of heat release rate. Figure 6.29 compares the ideal heat release rate controlled as the fuel pump flow rate and the actual heat release rate as measured by oxygen loss calorimetry. Initially at 500 kW the ideal and actual heat release rate are very close indicating that there is a very high combustion efficiency for the room with this relatively small heat release rate. As the heat release rate increases we see that the combustion efficiency decreases as less of the fuel being pumped into the room is actually burning which suggests that there should be large concentrations of hydrocarbons as well as intermediate species within the room. Figure 6.30 presents the temperature and species volume fractions inside the room and confirms that indeed there is a large quantity of hydrocarbons and other intermediate species in the room when the combustion efficiency is poor. Comparing the temperature plot with the heat release rate a maximum in temperature is observed at a measured heat release rate of ~600 kW with an ideal heat release rate of 750 kW. As the heat release rate is increased beyond this point the temperature in the room drops indicating that the fuel is acting as a diluent and cooling the room since there is no oxygen to consume it. Indeed at nearly the same time as the temperature peak in the heat release rate ramp the oxygen volume fraction drops to zero and the total hydrocarbons in the room begin to accumulate and larger volume fractions are measured. As the heat release rate is increased there is proportionately more hydrocarbons in the room.

Initially the CO volume fraction is increasing in the room with heat release rate. However the CO volume fraction also reaches a peak around 1400 s (~800 kW), this is a phenomenon that has been noticed by other researchers [54]. At the same time the CO_2 volume fraction is leveling off in the room. This indicates several things, first the volume fraction of CO in the room does not continue to rise indefinitely as the heat release rate rises. Also, the leveling off of CO_2 volume fraction indicates that beyond a certain point only a fixed volume of fuel is being consumed, likely related to a relatively steady state supply rate of O_2. The drop in CO production is likely related to the drop in temperature in the room as the lower temperature favors CO_2 production. In fact the water gas shift reaction indicates that CO would be consumed and more hydrogen would be produced as a byproduct. Figure 6.31 presents the GC measurements for the heat release rate test, ISOHept27. The data set is limited, however there does seem to be a slight increase in H_2 volume fraction corresponding to the decrease in CO volume fraction as predicted by the water gas shift reaction.

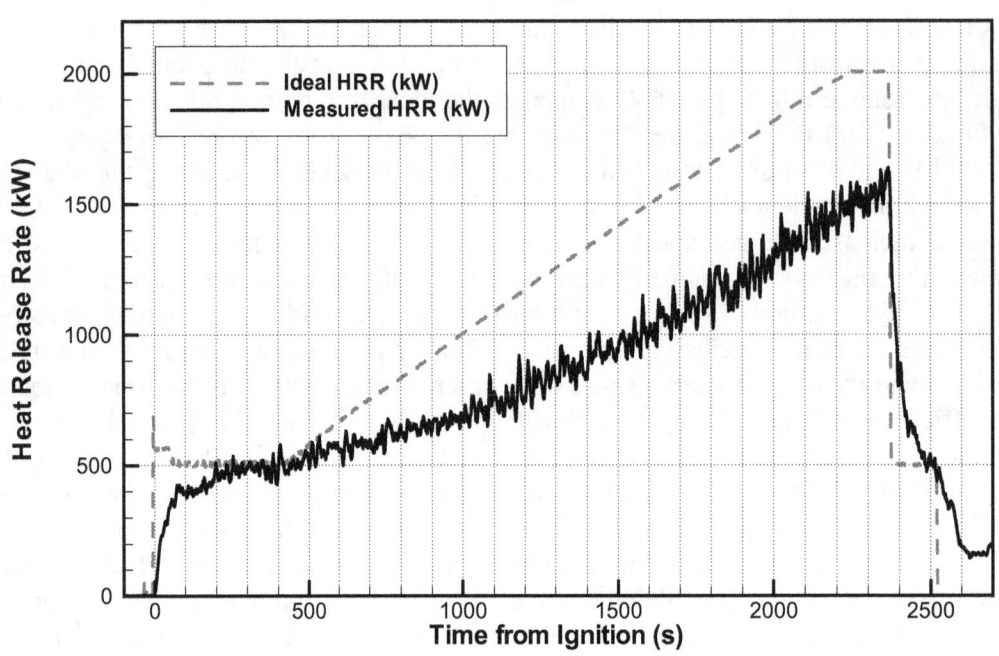

Figure 6.29: Comparison of ideal heat release rate, determined from fuel flow rate, and measured heat release rate measured from oxygen loss calorimetry for test ISOHept27, where the heat release rate was ramped linearly with time..

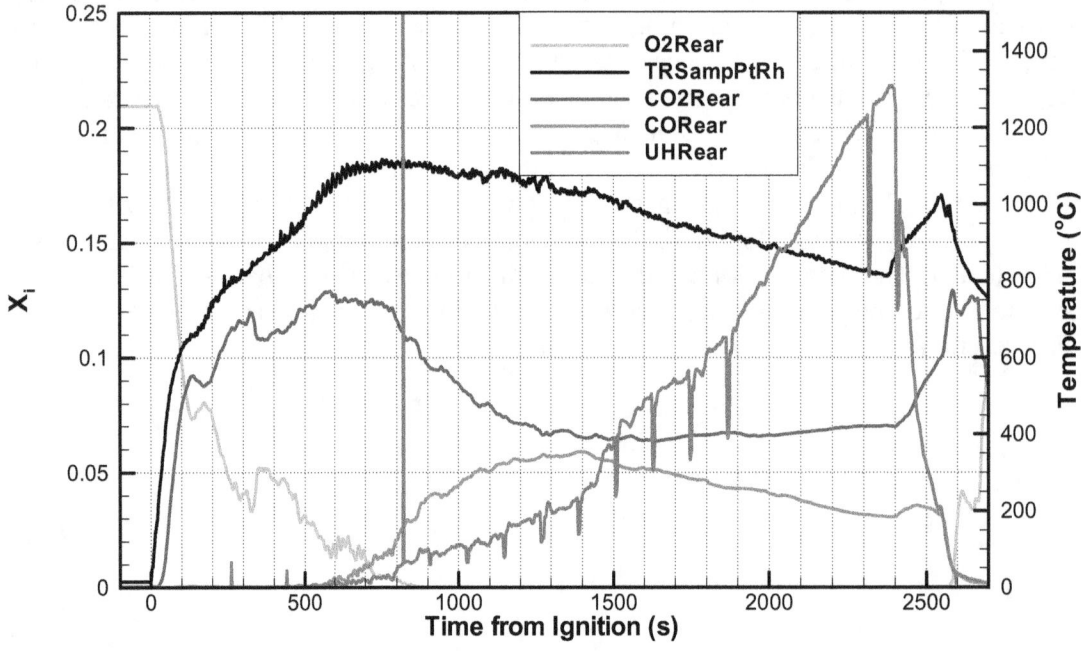

Figure 6.30: Comparison of temperature and various species mole fractions for ISOHept27 test with a linear heat release rate ramp.

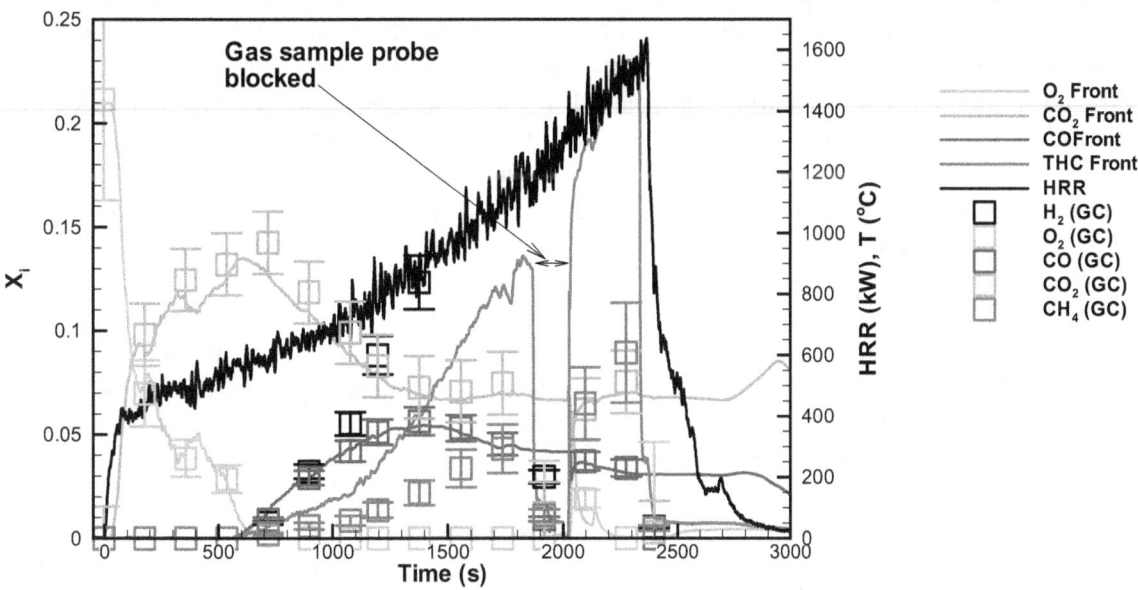

Figure 6.31: Comparison of various gas species from both gas analyzers and GC analysis with heat release rate for heat release rate ram test ISOHept27.

7 SUMMARY

This report documents a set of 30 full scale ISO 9705 room under-ventilated fire experiments looking at the effects of different fuels, different fuel distribution, ventilation effects, and how the fire changes as a function of heat release rate. The data for all of these experiments is collected and published in parallel with this report as a digital online resource that can be accessed at http://www.fire.nist.gov/testdata/FSE. Analysis was conducted on each case in order to verify the validity of results and to begin to explain some of the phenomena observed in the tests.

Gaseous, liquid, and solid fuels were all considered in order to provide a range of different fuels for numerical model validation but also to highlight the different behaviors of each fuel considered in an identical environment. Methane was the only gaseous fuel considered here with under-ventilated conditions reached for full door and quarter door experiments. A metered gravel burner was used to accurately measure the fuel supply rate. Heptane, iso-propanol, and toluene were the three liquid fuels considered here. For each liquid fuel a free burn pan burner was used in a variety of pan configurations in addition to a spray burner. Under-ventilated conditions were achieved for each fuel in each type of burner configuration. Nylon, polystyrene, and polypropylene are the solid fuels which were considered. Both single and distributed solid fuel sources were used. Each solid fuel was consumed in a free burn pan. Under-ventilated fires were achieved for both polystyrene and polypropylene when a total surface area of 1 m^2 was available for the fuel. Under-ventilated conditions were not achieved for nylon in this test series.

There were three forms of fuel delivery that were used in this investigation. First, and somewhat unsuccessfully, a water cooled pan burner that was pump fed was used for some of the early heptane fires. Second, and more predominately, free burn pans placed on digital mass load cells were used for both liquid and solid fuels. Thirdly a pump fed spray burner was used with the liquid fuels to provide longer duration steady state fire conditions than were achievable with the free burn pans. For the pump fed pan and the spray burners only one fuel distribution was utilized. Specifically, each burner was placed at the geometric center of the room. All of the pump fed pan experiments used a 1 m^2 pan with inclined walls so that it had a variable surface area which was difficult to measure. All of the spray burner experiments utilized a 0.5 m^2 pan attached to a load cell to catch the fuel spray and provide some indication if fuel was accumulating inside the room. A variety of different fuel distributions were utilized with the free burn pans. Three pan sizes with internal dimensions of 1 m^2 (100 cm by 100 cm), 0.5 m^2 (71c m by 71 cm), and 0.25 m^2 (50 cm by 50 cm) were used, each with a lip height of 0.1 m. Two load cells were used with the free burn pans to allow for either a single burner or two burners to be utilized. Distributed fuel source cases were investigated, comparing a fuel pan in the center of the room versus a fuel pan at the rear of the room versus two fuel pans, one located at the center of the room and one located at the rear of the room.

Enclosure ventilation was investigated with a heptanes spray burner. The doorway of the room was varied in width, and consequently area. Half (40 cm), quarter (20 cm), and eighth (10 cm) doorways were investigated. For the same heat release rate, similar temperatures were observed for each case, however dramatically different chemical compositions were observed. Specifically, as the doorway width was decreased the room became more under-ventilated. These results are helpful in testing and designing scaling relationships for enclosure fires.

The changes within the room as a function of heat release rate were illustrated using an enclosure with a one eighth width (10 cm) doorway and a heptanes spray burner. The heptanes fuel flow rate was incremented linearly with time to provide a linear increase in ideal heat release rate within the room. This allowed for an observation of how the thermal environment and the chemical environment of the room changes as a continuous function of heat release rate. This experiment allowed for the observation of maxima of both temperature and CO volume fraction in the room in the middle of the heat release rate ramp. Considering all of the fuels in this report a global peak in CO concentration was also observed in the hood after the combustion gases had left the room. The CO volume fraction peaked at approximately 1250 kW for all fuels.

A new real time extractive optical soot measurement probe was developed and tested with this set of experiments as well. The design of this probe is still in its infancy. However, a proof of concept was developed which illustrated that it is indeed possible to make a real time soot mass fraction measurement at an interior point within an enclosure during an under-ventilated compartment fire. Gravimetric soot data was also collected which features a lower uncertainty and provided the data for the optical soot probe calibration as well as a total carbon balance in the room.

Highly under-ventilated fire conditions were achieved for a number of fuels. The capture and measurement of carbon in all of its forms allowed for a detailed analysis of the compartment chemistry at discrete points within the room. This analysis was done in the form of mixture fraction state relationships. It was found that for $Z < Z_{st}$ the mixture fraction model does a good job of predicting the chemical composition within the room. For $Z > Z_{st}$ the agreement is not as good and in some cases really poor. This has been previously illustrated by other researchers.

However, by collecting real data that accounts for all carbon in the system at discrete points in a real fire, the foundation has been laid to develop better combustion models for fire researchers that still maintain the speed and simplicity of the mixture fraction model.

Scaling relationships allow for the comparison of data from this full scale ISO 9705 room with that obtained in the 2/5 scale compartment. Care must be taken when selecting scaling laws to compare compartments of different geometry even if they are only geometrically scaled. For example, the dimentionless compartment height showed a good agreement between the full and reduced scale experiments when the doorway width (ventilation) was scaled with the rest of the room. However, if the doorway width (ventilation) is not scaled with the volume of the room it may be necessary to apply two or more scaling relationships to understand the differences in the room. This was noted by the ventilation scaling factor which showed good agreement for change in vent size given a room of equal size, but needs to be used secondarily if both the room size and vent size are independently scaled.

As a whole this data provides a significant amount of information for modelers to use to validated their models for steady state conditions and to determine how well their models are able to recreate the effects of specific changes to the geometry, reactant supply, and fuel type. So far only fairly simple geometries, fuels, and fuel distributions have been utilized. Therefore it is important to note that this set of experiments and this report are only one step in a larger process. These results build on the reduced scale enclosure results [5] and will be expanded upon in the proposed investigations presented in section 8.

8 FUTURE WORK

These results build on the work conduced in the reduced scale enclosure [5] and provide a wealth of information on scaling of the enclosure, ventilation, and fuel heat release rate. However, these are generally global effects measured at a few discrete points that are generally well understood and may be relatively easy to model but may not be an adequate basis of validation for real and much more complicated scenarios that fire models are used to simulate. Therefore, in order to build on these results and expand the applicability and usefulness of this data, these simplistic experiments are going to be expanded to include more complicated and realistic fuel sources such as furniture.

It in planned that in the Spring of 2009 and 2010 another series of experiments will be conducted in which real fuel packages such as furniture will be burned and investigated. This will provide more realistic and complete validation data that is more convincing when you are trying to simulate a real fire situation. Adding real fuels to the existing suite of data will provide a robust set of conditions for modelers to validate against. This will allow, for example, an investigator who is having difficulty modeling a specific real fire scene to go back and look at the more global fire conditions and determine the specific source of the problem in a systematic way rather than randomly changing variables until the system works as desired.

The diagnostics used in this test will be expanded upon for the next set of experiments. It is planned to provide more detailed mapping of the interior of the ISO 9705 enclosure as well as some mapping of the doorway. This implementation will require the implementation of linear stages to move probes within the room and in the doorway.

9 REFERENCES

1. ISO9705, *Fire Tests - Full-Scale Room Test for Surface Products First Edition.* 1993, International Organization for Standardization: Geneva, Switzerland
2. McGrattan, K.B., et al., *Fire Dynamics Simulator (Version 5): Technical Reference Guide.* 2007, National Institute of Standards and Technology
3. National Fire Protection, A. and E. Society of Fire Protection, *SFPE handbook of fire protection engineering.* 2002, Quincy, Mass: National Fire Protection Association.
4. Floyd, J.E. and K.B. McGrattan. *Validation of A CFD Fire Model Using Two Step Combustion Chemistry Using the NIST Reduced-Scale Ventilation-Limited Compartment Data.* in *9th International IAFSS Symposium.* 2008. Karlsruhe, Germany.
5. Bundy, M., et al., *Measurements of Heat and Combustion Products in Reduced-Scale Ventilation-Limited Compartment Fires.* 2007, National Institute of Standrads and Technology.http://www.fire.nist.gov/testdata/RSE/
6. Beyler, C.L., *Major Species Production by Diffusion Flames in A 2-Layer Compartment Fire Environment.* Fire Safety Journal, 1986. **10**(1): p. 47-56.
7. Zukoski, E.E., et al., *Species Production and Heat Release Rates in 2-Layered Natural-Gas Fires.* Combustion and Flame, 1991. **83**(3-4): p. 325-332.
8. Cleary, M.J. and J.H. Kent, *Modeling of species in hood fires by conditional moment closure.* Combustion and Flame, 2005. **143 (4)**: p. 357-368.
9. Bryner, N.P., E.L. Johnson, and W.M. Pitts, *Carbon Monoxide Production in Compartment Fires - Reduced-Scale Enclosure Test Facility.* 1994, National Institute of Standards and Technology
10. Pitts, W.M., E.L. Johnsson, and N.P. Bryner, *Carbon Monoxide Formation in Fires By High-Temperature Anaerobic Wood Pyrolysis.* Twenty-Fifth Symposium (International) on Combustion, 1994: p. 1445-1462.
11. Lattimer, B.Y. and R.J. Roby, *Carbon monoxide levels in structure fires: Effects of wood in the upper layer at a post-flashover compartment fire.* Fire Technology, 1998. **34**(4): p. 325-355.
12. Gottuk, D.T., R.J. Roby, and C.L. Beyler, *The role of temperature on carbon monoxide production in compartment fires.* Fire Safety Journal, 1995. **24**(4): p. 315-331.
13. Lattimer, B.Y., U. Vandsburger, and R.J. Roby, *Species transport from post-flashover fires.* Fire Technology, 2005. **41**(4): p. 235-254.
14. Gann, R.G., et al., *Smoke component yields from room-scale fire tests.* 2003.http://fire.nist.gov/bfrlpubs/fire03/PDF/f03005.pdf
15. Hirschler, M.M., *Analysis of work on smoke component yields from room-scale fire tests.* Fire and Materials, 2005. **29**(5): p. 303-314.
16. Blomqvist, P. and A. Lonnermark, *Characterization of the combustion products in large-scale fire tests: Comparison of three experimental configurations.* Fire and Materials, 2001. **25**(2): p. 71-81.
17. Pitts, W.M., N.P. Bryner, and E.L. Johnson. *Combustion Product Formation in Under and Overventilated Full-Scale Enclosure Fires.* in *Combustion Institute/Central and Western States (USA) and Combustion Institute/Mexican National Section and American Flame Research Committee. Combustion Fundamentals and Applications. Joint Technical Meeting.* 1995. San Antonio, TX.
18. Bertin, G., J.M. Most, and M. Coutin, *Wall fire behavior in an under-ventilated room.* Fire Safety Journal, 2002. **37**(7): p. 615-630.

19. Snegirev, A.Y., et al., *Turbulent Diffusion Combustion under Conditions of Limited Ventilation: Flame Projection Through an Opening.* Combustion, Explosion and Shock Waves, 2003. **39**(1): p. 1-10.
20. Utiskul, Y., et al., *Compartment fire phenomena under limited ventilation.* Fire Safety Journal, 2005. **40**(4): p. 367-390.
21. Yii, E.H., A.H. Buchanan, and C.M. Fleischmann, *Simulating the effects of fuel type and geometry on post-flashover fire temperatures.* Fire Safety Journal, 2006. **41**(1): p. 62-75.
22. Yii, E.H., C.M. Fleischmann, and A.H. Buchanan, *Experimental study of fire compartment with door opening and roof opening.* Fire and Materials, 2005. **29**(5): p. 315-334.
23. Nist, *Final report on the collapse of the World Trade Center towers.* 2005, U.S. Dept. of Commerce, Technology Administration, National Institute of Standards and Technology: Gaithersburg, Md. http://wtc.nist.gov/pubs
24. Grosshandler, W.L., et al., *Report of the technical investigation of The Station nightclub fire.* 2005, U.S. Dept. of Commerce, Technology Administration, National Institute of Standards and Technology ; Washington, D.C. : For Sale by the Supt. of Docs., U.S. G.P.O.: Gaithersburg, Md.
25. Madrzykowski, D. and W.D. Walton, *Cook County Administration Building fire, 69 West Washington, Chicago, Illinois, October 17.* 2004
26. Hamins, A., et al., *Report of Experimental Results for the International Fire Model Benchmarking and Validation Exercise #3.* 2005
27. Huggett, C., *Estimation of Rate of Heat Release by Means of Oxygen-Consumption Measurements.* Fire and Materials, 1980. **4**(2): p. 61-65.
28. Parker, W.J., *Calculations of the Heat Release Rate by Oxygen-Consumption for Various Applications.* Journal of Fire Sciences, 1984. **2**(5): p. 380-395.
29. Bryant, R.A., et al., *The NIST 3 Megawatt Quantitative Heat Release Rate Facility - Description and Proceedure.* 2004
30. Siegel, R. and J.R. Howell, *Thermal Radiation Heat Transfer.* 3rd ed. 1992, New York: Hemisphere Publishing.
31. Choi, M.Y., et al., *Comparisons of the Soot Volume Fraction Using Gravimetric and Light Extinction Techniques.* Combustion and Flame, 1995. **102**(1-2): p. 161-169.
32. Dobbins, R.A., G.W. Mulholland, and N.P. Bryner, *Comparison of a Fractal Smoke Optics Model with Light Extinction Measurements.* Atmospheric Environment, 1994. **28**(5): p. 889-897.
33. e, A.E., *Standard Guide for Room Fire Experiments.* 2006
34. Blevins, L.G. and W.M. Pitts, *Modeling of bare and aspirated thermocouples in compartment fires.* Fire Safety Journal, 1999. **33**(4): p. 239-259.
35. Glawe, G.E., F.S. Simmons, and T.M. Stickney, *Radiation and Recovery Corrections and Time Constants of Several Chromel-Alumel Thermocouple Probe in High Temperature, High Velocity Gas Streams.* 1953
36. Bryant, R., et al., *Radiative heat flux measurement uncertainty.* Fire and Materials, 2003. **27**(5): p. 209-222.
37. Taylor, B.N. and C.E. Kuyatt, *Guidelines for Evaluating and Expressing the Uncertainty of NIST Measurement Results,* in *NIST Technical Note.* 1994, National Institute of Standards and Technology: Gaithersburg, MD

38. Brohez, S., *Uncertainty analysis of heat release rate measurement from oxygen consumption calorimetry.* Fire and Materials, 2005. **29**(6): p. 383-394.
39. Bilger, R.W., *Reaction-Rates in Diffusion Flames.* Combustion and Flame, 1977. **30**(3): p. 277-284.
40. Peters, N., *Laminar Diffusion Flamelet Models in Non-Premixed Turbulent Combustion.* Progress in Energy and Combustion Science, 1984. **10**(3): p. 319-339.
41. Hamins, A. and K. Seshadri, *The Structure of Diffusion Flames Burning Pure, Binary, and Ternary Solutions of Methanol, Hepatane, and Toluene.* Combustion and Flame, 1987. **68**(3): p. 295-307.
42. Sivathanu, Y.R. and G.M. Faeth, *Generalized State Relationships for Scalar Properties in Nonpremixed Hydrocarbon Air Flames.* Combustion and Flame, 1990. **82**(2): p. 211-230.
43. Floyd, J.E., C.J. Wieczorek, and U. Vandsburger, *Simulation of the Virginia Tech Fire Research Laboratory Using Large Eddy-Simulation With Mixture Fraction Chemistry and Finite Volume Radiative Heat Transfer.* Proceedings of the 9th International Interflam Conference, Volume 1.September 17-19, 2001, Edinburgh, Scotland, 2001: p. 767-778.
44. Pitts, W.M., *The Global Equivalence Ratio Concept and the Formation Mechanisms of Carbon-Monoxide in Enclosure Fires.* Progress in Energy and Combustion Science, 1995. **21**(3): p. 197-237.
45. Sivathanu, Y.R. and G.M. Faeth, *Temperature Soot Volume Fraction Correlations in the Fuel-Rich Region of Buoyant Turbulent-Diffusion Flames.* Combustion and Flame, 1990. **81**(2): p. 150-165.
46. Koylu, U.O. and G.M. Faeth, *Carbon-Monoxide and Soot Emissions from Liquid-Fueled Buoyant Turbulent-Diffusion Flames.* Combustion and Flame, 1991. **87**(1): p. 61-76.
47. Puri, R. and R.J. Santoro, Proc.3rd Int.Sym.Fire Safety Science, 1991: p. 595-604.
48. Tewarson, A., F.H. Jiang, and T. Morikawa, *Ventilation-Controlled Combustion of Polymers.* Combustion and Flame, 1993. **95**(1-2): p. 151-169.
49. McCaffrey, B.J. and G. Heskestad, *Robust Bidirectional Low-Velocity Probe for Flame and Fire Applications.* Combustion and Flame, 1976. **1** p. 125-127.
50. Quintiere, J.G., *Scaling Applications in Fire Research.* Fire Safety Journal, 1989. **15**(1): p. 3-29.
51. Hull, T.R., J.M. Carman, and D.A. Purser, *Prediction of CO evolution from small-scale polymer fires.* Polymer International, 2000. **49**(10): p. 1259-1265.
52. Pitts, W.M., *Toxic Yield, Technical Basis for Performance Based Fire Regulations.* A Discussion of Capabilities, Needs and Benefits of Fire Safety Engineering.United Engineering Foundation Conference.Proceedings., 2001: p. 76-87.
53. Lock, A., et al. *Measurements in Standard Room Scale Fires.* in *9th International IAFSS Symposium.* 2008. Karlsruhe, Germany.
54. Ukleja, S. and M.A. Delichatsios, *Carbon monoxide and smoke production downstream of a compartment for underventilated fires.* Proceedings of the 9th International IAFSS Symposium, .September 21-26, 2001, Karlsruhe, Germany, 2008.

10 ACKNOWLEDGEMENTS

The authors wish to acknowledge the contributions of several people without whom this work would not be possible. The large fire lab staff, Lauren Delauder, Doris Reinhart, Tony Chakalis, and Greg Masenheimer did an excellent job of preparing and running the experiments as well as providing technical services to facilitate this test matrix. Kevin McGrattan and Jason Floyd provided invaluable insight and guidance into the design of the test matrix used here. Kelly Opert was a SURF student who provided help with developing plots and contributed to the post processing of data.

APPENDICES

A. CHANNEL LISTS

Testra							#	Testra					
Surface Temp Array Total 48 Gauge Center Ceiling E PtRh Bare Bead							71	TSBCCEPtRh	Center	3	6	C	TC	100
Interior Surface Temp Side Wall Adjacent to Front Sampling Position							72	TSFSamp	Center	3	7	C	TC	100
Exterior Surface Temp Side Wall Adjacent to Front Sampling Position							73	TSFSamp	Center	3	8	C	TC	100
Liquid Burner Fuel Temperature	119.5	178					74	TLiq	Center	3	10	C	TC	100
Liquid Burner Fuel Column Height	119.5	178		Level with burner floor surface (pool height =0) B13			75	HFuel	Center	3	11	cm	Ca	1
Liquid Burner Coolant Temperature Inlet				Connected to base of burner for pool height B14			76	TCoolIn	Center	3	12	C	TC	100
Liquid Burner Coolant Temperature Outlet				outside room water inlet B11			77	TCoolOut	Center	3	13	C	TC	100
TC Tree Rear TC 1 in up	120	288	2.5	inside burner in 3/8" water outlet tube B12			78	TR3	Center	3	14	C	TC	100
TC Tree Rear TC 1 ft up	120	288	30	Rear 1 ft up			79	TR30	Center	3	15	C	TC	100
TC Tree Rear TC 2 ft up	120	288	60	Rear 2 ft up			80	TR60	Center	3	16	C	TC	100
TC Tree Rear TC 3 ft up	120	288	90	Rear 3 ft up			81	TR90	Center	3	17	C	TC	100
TC Tree Rear TC 3.5 ft up	120	288	105	Rear 3.5 ft up			82	TR105	Center	3	18	C	TC	100
TC Tree Rear TC 4 ft up	120	288	120	Rear 4 ft up			83	TR120	Center	3	19	C	TC	100
TC Tree Rear TC 4.5 ft up	120	288	135	Rear 4.5 ft up			84	TR135	Center	3	20	C	TC	100
TC Tree Rear TC 5 ft up	120	288	150	Rear 5 ft up			85	TR150	Center	3	21	C	TC	100
TC Tree Rear TC 6 ft up	120	288	180	Rear 6 ft up			86	TR180	Center	3	22	C	TC	100
TC Tree Rear TC 7 ft up	120	288	210	Rear 7 ft up			87	TR210	Center	3	23	C	TC	100
TC Tree Rear TC 7 ft 11 in up	120	288	237.5	Rear 7 ft 11 in up			88	TR237	Center	3	24	C	TC	100
TC Tree Front TC 1 in up	120	72	2.5	Front 1 in up			89	TF3	Center	3	25	C	TC	100
TC Tree Front TC 1 ft up	120	72	30	Front 1 ft up			90	TF30	Center	3	26	C	TC	100
TC Tree Front TC 2 ft up	120	72	60	Front 2 ft up			91	TF60	Center	3	27	C	TC	100
TC Tree Front TC 3 ft up	120	72	90	Front 3 ft up			92	TF90	Center	3	28	C	TC	100
TC Tree Front TC 3.5 ft up	120	72	105	Front 3.5 ft up			93	TF105	Center	3	29	C	TC	100
TC Tree Front TC 4 ft up	120	72	120	Front 4 ft up			94	TF120	Center	3	30	C	TC	100
TC Tree Front TC 4.5 ft up	120	72	135	Front 4.5 ft up			95	TF135	Center	3	31	C	TC	100
TC Tree Front TC 5 ft up	120	72	150	Front 5 ft up			96	TF150	Center	4	0	C	TC	100
TC Tree Front TC 6 ft up	120	72	180	Front 6 ft up			97	TF180	Center	4	1	C	TC	100
TC Tree Front TC 7 ft up	120	72	210	Front 7 ft up			98	TF210	Center	4	2	C	TC	100
TC Tree Front TC 7 ft 11 in up	120	72	237.5	Front 7 ft 11 in up			99	TF237	Center	4	3	C	TC	100
Interior Surface Temperature Back Wall Centerflow Top	125	356	180	4,5			100	TSBWCTop	Center	4	4	C	TC	100
Interior Surface Temperature Back Wall Centerflow Middle	125	356	120	4,7			101	TSBWCMid	Center	4	5	C	TC	100
Interior Surface Temperature Back Wall Centerflow Bottom	125	356	60	4,9			102	TSBWCBot	Center	4	6	C	TC	100
Mass Flow Controller Output Rack 1				inside rack #1			103	MFRack1	Center	4	7	V	Ca	1
Mass Flow Controller Output Rack 2				inside rack #2			104	MFRack2	Center	4	8	V	Ca	1
Mass Flow Controller Output Rack 3				inside rack #3			105	MFRack3	Center	4	9	V	Ca	1
Exterior Surface Temperature Back Wall Centerflow Top	125	360	180	4,6			106	TSNEWCTop	Center	4	10	C	TC	100
Exterior Surface Temperature Back Wall Centerflow Middle	125	360	120	4,8			107	TSNEWCMid	Center	4	11	C	TC	100
Exterior Surface Temperature Back Wall Centerflow Bottom	125	360	60	4,10			108	TSNEWCBot	Center	4	12	C	TC	100
Vertical Stage Temperature 1							109	TVertStage1	Center	4	13	C	TC	100
Vertical Stage Temperature 2							110	TVertStage2	Center	4	14	C	TC	100
Vertical Stage Temperature 3							111	TVertStage3	Center	4	15	C	TC	100
Neat Soot Probe Laser Detector #1 (variation 0-10V)							112	SootDet1Var	Center	4	16	V	Ca	1
Neat Soot Probe Laser Detector #2 signal 0-10V)							113	SootDet2Sig	Center	4	17	V	Ca	1
Neat Soot Probe Cooling Water In Temp							114	TSootWaterIn	Center	4	18	C	TC	100
Neat Soot Probe Cooling Water Out Temp							115	TSootWaterOut	Center	4	19	C	TC	100
Neat Soot Probe Laser Detector #1 (variation Temp)							116	TSootFuel1Var	Center	4	20	C	TC	100
Neat Soot Probe Laser Detector #2 signal Temp)							117	TSootFuel2Sig	Center	4	21	C	TC	100
Neat Soot Probe Sample Temperature At Measurement							118	TSootSampR	Center	4	22	C	TC	100
Aspirated Pump 1 Pressure (TRSampA and TFTreeTopA)				outside RSE			119	PPump1	Center	4	23	Pa	Ca	1
Aspirated Pump 2 Pressure (TRTreeTopA)				outside RSE			120	PPump2	Center	4	24	Pa	Ca	1
Doorway Pressure Top							121	PDTop	Center	4	25	Pa	Ca	1
Doorway Pressure Bottom							122	PDBot	Center	4	26	Pa	Ca	1
Doorway Top Aspirated Temperature							123	TDTopA	Center	4	27	C	TC	100
Doorway Top Bare Bead BB6 Temperature							124	TDTopBBBB	Center	4	28	C	TC	100
Aspirated Pump 1 Temperature (TRSampA and TFSampA)				B1			125	TPump1	Center	4	29	C	TC	100
Aspirated Pump 2 Temperature (TRTreeTopA)				B2			126	TPump2	Center	4	30	C	TC	100
Extra Voltage 1							127	VX1	Center	4	31	V	Ca	1
Extra Voltage 2							128	VX2	Center	5	0	V	Ca	1
Extra Voltage 3							129	VX3	Center	5	1	V	Ca	1
4 V Marker Channel							130	5VMarker	Center	5	2	V	Ca	1
	Created Channels													
Event Marker 1							131	Event1	Center					
Event Marker 2							132	Event2	Center					
Doorway Velocity Top							133	VDTop	Center	B	0	m/s		n/a
Doorway Velocity Bottom							134	VDBot	Center	B	1	m/s		n/a
Soot Ratio							135	SootRatio	Center	B	2	n/a		n/a
not used							136	tbd	Center	B	3	n/a		n/a
not used							137	tbd	Center	B	4	n/a		n/a
not used							138	tbd	Center	B	5	n/a		n/a
not used							139	tbd	Center	B	6	n/a		n/a
not used							140	tbd	Center	B	7	n/a		n/a
not used							141	tbd	Center	B	8	n/a		n/a
not used							142	Efs	Center	C	0	V/V		n/a
not used							143	not used	Center	C	1	n/a		n/a
Soot Volume Fraction from Optical Measurement							144	Fv	Center	C	2	n/a		n/a
not used							145	not used	Center	C	3	n/a		n/a

B. EQUIPMENT LIST

Description	Manufacturer	Model	Serial#	NIST#
Oxygen analyzer for HRR	Servomex	540A		549709
CO_2/CO analyzer for HRR	Seimens	Ultramat 6		615207
Total HC analyzer for HRR	Rosemount	400A		569041
Mass flow controller for HRR	MKS	1179A53C	000346712	
Dew Point Transmitter for HRR	Vaisala	DMT242	A4850006	
Sample dryer for HRR	PermaPure	PD-200T-72SS	973-0905-6	
Micro GC for natural gas	Agilent	3000A		623489
Sample pump for HRR	Gast	MOA-P122-AA	4Z026	
Liquid fuel turbine flow meter	Exact Flow			
Natural gas flow meter	Instromet	IRMA 15M-125	319396	605032
Total heat flux gauge (HF Front)	Medtherm	16-0.75-10-4-12-36-20679k	131836	
Total heat flux gauge (HF Rear)	Medtherm	16-0.75-10-4-12-36-20679k	131835	
Oxygen Analyzer (O2Rear)	Servomex	4100	393063	623487
Oxygen Analyzer (O2Front)	Servomex	4100	393064	623488
CO_2/CO Analyzer (Rear)	Seimens	Ultramat 6E	NI-L00197	600671
CO_2/CO Analyzer (Front)	Seimens	Ultramat 6E		609425
Total HC analyzer (Rear, Front, and Movable)	Baseline-Mocon	8800 H		625764, 623892
Dew point meter (Rear, and Front)	Vaisala	DMT242	B074008, B074009	
Mass flow controller (Rear&Front)	MKS	M100B53C		
MFC power supply	MKS	247D		
Gas Chromatograph (Front)	Agilent	3000A	US10713004	
Pressure Transducer for Velocity	MKS	220DD		
Flow meter (spot check flows)	Bios Dry Cal	DCLT 20K		
Sample pump for aspirated TCs and gas sample tests #1-6	Gast	DOA-P703-FB		
Gas conditioning system (Rear&Front)	PermaPure	MG-2812		rental
Glass-lined stainless steel tubing	Grace Davison	3149		
Soot sample MFC	MKS	M100B53CCS1BV	021407828	
Soot sample MFC	MKS	M100B53CCS1BV	021407829	
MFC power supply	MKS	247D	000763015	
Soot sample filter	Pall	P5PJ047		
Soot sample cleaning pad	Hoppe's	1203		
Soot sample filter holder	Gelman Sciences	2220		
Soot sample 3-way solenoid valves	Parker	04F30C2208AAF4C05		
Soot sample pumps	Gast	MOA-P122-AA		

C. MICRO-GC METHOD REPORT

Method Report

Method:	March_iso
Method last saved:	Friday, May 09, 2008 11:25:49 AM
Method note:	
GC Description:	4 Channel Micro-GC

Method Miscellaneous Settings

No Method instructions file defined

No Pre-Run Program

Post-Run Program:
    [%INST_TYPE%] [%INST_SN%]
    Wait for completion: Off

Report Settings:
Type(s):
    ESTD Report

Output(s):
    Text file: C:\Program Files\Agilent\Cerity QA-QC\DefaultDataDirectory\Default%06NEXT%.txt
    HTML file: C:\Program Files\Agilent\Cerity QA-QC\DefaultDataDirectory\Default%06NEXT%.htm

Output(s) Generated:
    At End of Run

3000 GC Setpoints	A	B	C	D
Sample Inlet Temperature (°C)	100 [ON]	Same as A	Same as A	Same as A
Injector Temperature (°C)	100 [ON]	100 [ON]	100 [ON]	100 [ON]
Column Temperature (°C)	100 [ON]	110 [ON]	90 [ON]	65 [ON]
Sampling Time (s)	15 [ON]	15 [ON]	Same as B	Same as B
Inject Time (ms)	0	30	25	25
Run Time (s)	120	120	120	120
Post Run Time (s)	0	0	0	0
Pressure Equilibration Time (s)	0	10	10	10
Column Pressure (psi)	40.00 [ON]	25.00 [ON]	25.00 [ON]	25.00 [ON]
Post Run Pressure (psi)	40.00 [ON]	25.00 [ON]	25.00 [ON]	25.00 [ON]
Detector Filament	Enabled	Enabled	Enabled	Enabled
Detector Sensitivity	High	High	High	High
Detector Data Rate (Hz)	100	100	100	100
Baseline Offset (mV)	0	0	0	0
Backflush Time (s)	5.0	3.0	n/a	n/a

3000 GC Configuration	A	B	C	D
Injector Type	Backflush	Backflush	Timed	Timed
Carrier Gas	Argon	Helium	Helium	Helium
Column Type	Molecular Sieve	Plot U	OV-1	Stabilwax
Detector Type	TCD	TCD	TCD	TCD
Inlet Type	Heated	Heated	Heated	Heated

```
Integrator Settings and Timed Events:  Signal   1
```

Initial Setting	Value
Slope Sensitivity	7000.000
Peak Width	0.020
Area Reject	1.000
Height Reject	1.000
Shoulders	OFF
Advanced Baseline	OFF

Time	Event	Value
0.000	Integration	OFF
0.500	Integration	ON

```
Integrator Settings and Timed Events:  Signal   2
```

Initial Setting	Value
Slope Sensitivity	5000.000
Peak Width	0.020
Area Reject	1.000
Height Reject	1.000
Shoulders	OFF
Advanced Baseline	OFF

Time	Event	Value
0.000	Integration	OFF
0.550	Integration	ON

```
Integrator Settings and Timed Events:  Signal   3
```

Initial Setting	Value
Slope Sensitivity	500.000
Peak Width	0.040
Area Reject	1.000
Height Reject	1.000
Shoulders	OFF
Advanced Baseline	OFF

Time	Event	Value
0.000	Integration	OFF
0.300	Integration	ON

```
Integrator Settings and Timed Events:  Signal   4
```

Initial Setting	Value
Slope Sensitivity	5000.000
Peak Width	0.040
Area Reject	1.000
Height Reject	1.000
Shoulders	OFF
Advanced Baseline	OFF

Time	Event	Value
0.000	Integration	OFF
0.500	Integration	ON

Calibration Table

Calculation Base	Peak Area
Rel. Reference Window	5.000 %
Abs. Reference Window	0.000 min
Rel. Non-ref. Window	5.000 %
Abs. Non-ref. Window	0.000 min
Sample Defaults	
Amount	100.0000
Units	mole%
Multiplier	1.0000
Dilution	1.0000
Curve Type	Piecewise
Origin	Included
Weight	Equal
Recalibration Settings	
Correct All Ret. Times	No, only for identified peaks
Partial Calibration	Yes, identified peaks are recalibrated
Average Response	Floating Average New 100%
Average Retention Time	Floating Average New 100%
Signal 1	TCD1 A
Uncalibrated Peaks: do not quantify	
Signal 2	TCD2 A
Uncalibrated Peaks: do not quantify	
Signal 3	TCD3 A
Uncalibrated Peaks: do not quantify	
Signal 4	TCD4 A
Uncalibrated Peaks: do not quantify	

Retention Time [min]	Sig	Lvl	Amount	Area	Amt/Area	Ref	Multiplier	Peak Name
0.350	3	1	1.34000	10000.00000	0.00013		1.00000	H2O

		3	0.67000	4700.00000	0.00014		H2O
0.425	3	1	0.10000	2039.50000	0.00005	1.00000	1-Butene
0.429	3	1	5.97400	237376.80000	0.00003	1.00000	n-Butane
		2	0.05010	1148.00000	0.00004		n-Butane
		3	0.10000	1.00000	0.10000		n-Butane
0.471	3	1	3.00000	32074.79195	0.00009	1.00000	C4H6
0.520	4	1	0.00000	716.70200	0.00000	1.00000	Peak@0.52
0.561	3	1	1.85400	19507.70650	0.00010	1.00000	C4H8
0.583	1	1	12.47200	63377.00000	0.00020	1.00000	H2
		2	0.09860	475.05000	0.00021		H2
0.589	3	1	0.10000	1920.40000	0.00005	1.00000	1-Pentene
0.617	3	1	0.10020	1801.62000	0.00006	1.00000	n-Pentane
0.638	2	1	2.99000	60411.00000	0.00005	1.00000	CO2
		2	8.99900	147673.30000	0.00006		CO2
		3	0.04990	1072.84000	0.00005		CO2
0.668	3	1	1.00000	7057.55276	0.00014	1.00000	Pentane
0.690	2	1	1.99600	40519.00000	0.00005	1.00000	Ethylene
		2	0.04930	1081.12000	0.00005		Ethylene
		3	0.10000	1614.80000	0.00006		Ethylene
0.691	4	1	6.00000	110000.00000	0.00005	1.00000	CH3OH
		2	1.19000	27200.00000	0.00004		CH3OH
0.700	1	1	20.95000	11479.00000	0.00183	1.00000	O2
		2	0.05000	33.80000	0.00148		O2
		3	16.00000	9010.00000	0.00178		O2
		4	8.00000	4537.70000	0.00176		O2
0.729	2	1	3.99600	85260.00000	0.00005	1.00000	Ethane
		2	0.05480	1364.20000	0.00004		Ethane
		3	0.10000	1.00000	0.10000		Ethane
0.827	1	1	37.28700	16577.00000	0.00225	1.00000	N2
		3	79.05000	36719.00000	0.00215		N2
		4	100.00000	46143.53000	0.00217		N2
		5	0.10000	91.74000	0.00109		N2
0.832	2	1	0.99400	17009.00000	0.00006	1.00000	Acetylene
		2	0.05010	905.93000	0.00006		Acetylene
0.919	3	1	0.10000	1967.00000	0.00005	1.00000	1-Hexene
0.983	3	1	0.09900	2240.03957	0.00004	1.00000	n-Hexane
		2	0.04990	1.00000	0.04990		n-Hexane
1.100	1	1	5.00100	5067.00000	0.00099	1.00000	CH4
		2	99.00000	104201.30000	0.00095		CH4
		3	0.10030	1.00000	0.10030		CH4
1.355	3	1	0.20000	553.00000	0.00036	1.00000	Hexanes-AJL
1.390	1	1	1.00700	419.00000	0.00240	1.00000	CO
		2	3.99500	1714.45000	0.00233		CO
1.435	3	3	0.00000	4526.97558	0.00000	1.00000	Heptanes@1.435
1.519	3	3	0.00000	10637.99074	0.00000	1.00000	Heptanes@1.519
1.547	4	1	0.00000	2649.07084	0.00000	1.00000	Heptanes@1.547
1.623	3	3	0.00000	1832.17510	0.00000	1.00000	Heptanes@1.623
1.670	3	1	1.00000	22.00000	0.04545	1.00000	CH2O
1.742	3	3	1.64000	43200.00000	0.00004	1.00000	n-Heptane
		4	0.04990	1018.00000	0.00005		n-Heptane

INDEX

Burners 9
   Locations 10
Calorimetery 11
Ceramic Insulation Retainers 8
Delay Times 13, 15, 33
Doorway 9
Enclosure
   Full Scale Enclosure (FSE) 4
   Reduced Scale Enclosure (RSE) 2, 3
equivalence ratio 111
FDS **1**
Fuel Distribution 125
Fuel Properties 125
fuel types 4, 5
Fuel Types
   Liquid Fuels 114
   Solid Fuels 120
Gas Analyzers 13
Gas Chromatography
   Calibration Standards 17
   Columns 17
Gas Sample Storage System 19
Gas Species Measurements
   GC vs. Gas Analyzer 65
   Wet/Dry Basis 65
Global equivalence ratio (GER) 2
Gravimetric Soot Sampling 21
Heat flux 29
Heat Release Rate
   normalized 110
Hood 11
HRR
   Distributed Fuels 41
   Gaseous Fuels 40
   Liquid Fuels 41
   Ramp 42
   Solid Fuels 41
   Spray Burner 42
   Steady States 42
ISO 9705 .... 4, 6
Labview 33
Load cell 9

Matlab 33
NIST Large Fire Research Laboratory (LFRL) 11
Real Time Extractive Soot Measurement . 23
Reduced Scale Enclosure (RSE) 3
Response Time 13
Sample Probe Locations 7
Sample probes
   Gas Drying 14
   Locations 32
Sample Probes 14
Scaling 110
   Doorway 131
   Ventilation 112
Soot
   Filtering 14
Spray Nozzle 9
stoichiometric ratio 111, 125
Test Conditions 38
Thermocouples 25
   Aspirated Thermocouples 25
   Compare Bare Bead and Aspirated 50
   Radiation Effects 28
   Repeatablility 50
Total Hydrocarbon Analyzer 13
Uncertainty **35**
   Burner Dimensions 9
   Ceramic Fiber Blanket Composition 6
   Data Acquisition (DAQ) 33
   Doorway Width 9
   Gas Chromatography 18
   Heat Flux 30
   Listing 35
   Load Cell 9
   Mass Balance 21
   Room Dimensions 6
   Soot Measurements 21
   Span gas 14
   Thermocouple, Aspirated 26
   Thermocouple, Bare Bead 29, 50
   Water Correction 34
Validation

RSE 2	Venturi Pump 26, 28
Ventilation 130	Water correction 33

www.ingramcontent.com/pod-product-compliance
Lightning Source LLC
Chambersburg PA
CBHW080249180526
45167CB00006B/2465